Multiple Choice Questions in
PHYSICS

Multiple Choice Questions in
PHYSICS

S. MOHAN, Ph.D, D.Sc.,

Former Vice-Chancellor, PRIST University, Thanjavur
Senior Professor of Materials Science
Hawassa University, Hawassa, Ethiopia
Dean-Research, Vel Tech University,
Avadi, Chennai

Chennai   Trichy   NewDelhi

ISBN 978-81-8094-292-1 **MJP Publishers**

All rights reserved No. 44, Nallathambi Street,
Triplicane,
Chennai 600 005

MJP 032 © Publishers, 2020

Publisher : C. Janarthanan

This book has been published in good faith that the work of the author is original. All efforts have been taken to make the material error-free. However, the author and publisher disclaim responsibility for any inadvertent errors.

Preface

Most competitive examinations adopt multiple-choice tests. Virtually everyone is familiar with this type of test and understands how to proceed. In order to succeed University Examinations, GATE, NET and SLET examinations, students will need both specific knowledge of the subject matter and test-taking skills and abilities. Even with some background in physics through self-study, leisure reading and precious coursework, it is unlikely that one can simply proceed to take and pass the examination. They will need to review the basics and the sample materials. They should also have to read several textbooks recommended at the University level. Succeeding in examinations requires planning and preparation. The competitive examinations in multiple-choice questions are designed to allow students to demonstrate the knowledge you have gained in a variety of topics in physics. They are intended to help you to meet a wide diversity of personal and professional goals.

This book will help students to prepare for the University Examinations, GATE, NET, and SLET examinations. It also provides basic information about each topic in physics.

The following suggestions will help one to achieve success in the objective questions.

1. Manage your time efficiently. You should know how much time you are given for an average question. The time you need for each individual question, of course, will vary.

2. Postpone answers to tough questions. Do not "hang up" on any one question. If it is difficult, go to the next question but mark it so that you can return to it later. You should not spend much time on questions you know very little about.

3. Always give an answer to every question. You can only gain if you answer a question. If you have no other basis, make a guess.

4. Mark your answers clearly and neatly. Your answer sheet is scored by an optical scanner. To do this accurately, the machine needs clearly marked answers. In particular, if you change an answer, erase THOROUGHLY.

5. Consider all possibilities. Remember the multiple choice questions work because at least some of the WRONG answers appear to be correct. Do not base your final answer on a reading of just one alternative. Read the entire set and form a final judgment after considering all of the possibilities.

6. Avoid clerical errors in recording your answers. It is possible to inadvertently "skip" a question and proceed to mark answers in the wrong locations. Verify that the question number and the answer number are the same in each case.

Wishing you good luck in your venture!

S Mohan

Contents

1. Classical Mechanics — 1
2. Statistical Physics — 45
3. Mathematical Physics — 77
4. Solid State Physics — 109
5. Quantum Mechanics — 149
6. Electromagnetic Theory and Electronics — 203
7. Atomic and Molecular Spectroscopy — 283
8. Nuclear Physics and Elementary Particles — 313

Test paper 1 — 347

Test paper 2 — 363

Test paper 3 — 379

1

CLASSICAL MECHANICS

FORMULAE

1. If a force F is such that the work W done around a closed curve is zero, i.e., $\oint F.dr = 0$, then the force and system are said to be conservative or, if $\nabla \times F = 0$, then the force and system are said to be conservative.

2. If a system of N particles is moving independently of each other, then the number of degrees of freedom is $3N$.

3. The equation of constraint for a particle moving on or outside the surface of a sphere of radius a is $x^2 + y^2 + z^2 \geq a^2$, if the origin of the coordinate system coincides with the centre of the sphere.

4. The general form of holonomic constraint is
$$\phi_j(q_1, q_2, ..., q_n, t) = 0 \qquad (j = 1, 2, ..., k)$$
Here, $q_1, q_2, ..., q_n$ are the generalized coordinates.

5. Consider the motion of two particles in the xy plane. If these particles are connected by a rigid rod of length l then the corresponding equation of constraint is
$$(x_2 - x_1)^2 + (y_2 - y_1)^2 - l^2 = 0$$

6. The general form of non-holonomic constraint is
$$\sum_{i=1}^{n} a_{ji} dq_i + a_{jt} dt = 0 \qquad (j = 1, 2, ..., m)$$
where, a's are, in general functions of the q's and t.

7. A system consisting of N particles, free from constraints, has $3N$ independent coordinates or degrees of freedom. If the sum of degrees of freedom of all the particles is K, then the system may be regarded as a collection of free particles subjected to $(3N - K)$ independent constraints.

8. The generalized velocities of a system are

$$\dot{q}_i = \frac{dq_i}{dt} \quad (i = 1, 2, 3, ..., k)$$

Here, $\dot{q}_i$ is the generalized coordinate.

9. The virtual work of the constraint forces is

$$\delta W_C = \sum_{i=1}^{N} F_i \cdot \delta r_i$$

Here,

F_i = constraint force
δr_i = the virtual displacement of the ith particle of the given system

10. The principle of virtual work is

$$\delta W = \sum_i F_i^a \cdot \delta r_i = 0$$

Here, F_i^a = applied force.

11. The mathematical statement of D'Alembert's principle is

$$\sum_i (F_i - p_i) \cdot \delta r_i = 0$$

Here,

F_i = applied force
p_i = addition force
$(F_i - p_i)$ is known as "reversed effective force" on ith particle.

12. Lagrangian function is given by

$$L(q, \dot{q}, t) = T(q, \dot{q}, t) - V(q, t)$$

Here,

q = generalized coordinate
$\dot{q}$ = generalized velocity

T = kinetic energy
V = potential energy

For a conservative system, the Lagrangian function is

$$L(q, \dot{q}) = T(q, \dot{q}) - V(q)$$

13. The generalized momentum p_i conjugate to the generalized coordinate q_i is

$$p_i = \frac{\partial L}{\partial \dot{q}_i}$$

14. If the motion of the holonomic system is described by the generalized coordinates $q_1, q_2,..., q_k$ and the generalized velocities $\dot{q}_1, \dot{q}_2,..., \dot{q}_k$ then the equation of motion is

$$\frac{d}{dt}\left(\frac{\partial T}{\partial \dot{q}_i}\right) - \frac{\partial T}{\partial q_i} = Q_i \quad (i = 1, 2,..., k)$$

Here,
 T = kinetic energy of the system
 Q_i = generalized force

15. For conservative systems, Lagrange's equations can be written in the form

$$\frac{d}{dt}\left(\frac{\partial L}{\partial \dot{q}_i}\right) - \frac{\partial L}{\partial q_i} = 0$$

16. The Lagrange's equation for the simple pendulum is

$$\ddot{\theta} + \frac{g}{l}\sin\theta = 0$$

Here,
 l = length of the pendulum
 g = acceleration due to gravity

17. The Hamiltonian function of a holonomic system having k degrees of freedom is the function of the generalized coordinates and momenta of the system. It is represented as

$$H(q, p, t) = \sum_{i=1}^{K} p_i \dot{q}_i - L = T + V$$

18. For a particle of mass m in a conservative field $V(x, y, z)$, the Hamiltonian is

$$H = \frac{1}{2m}(p_x^2 + p_y^2 + p_z^2) + V(x, y, z)$$

Here,

$\dfrac{p_x^2 + p_y^2 + p_z^2}{2m}$ = kinetic energy

$V(x, y, z)$ = potential energy

19. Hamilton's canonical equations of motion for a holonomic system with k degrees of freedom, subject only to conservative forces, are

$$\dot{q}_i = \frac{\partial H}{\partial p_i}; \quad \dot{p}_i = -\frac{\partial H}{\partial q_i} \quad (i = 1, 2, 3, ..., k)$$

20. The kinetic energy in terms of generalized coordinates $q_j (j = 1, 2, 3, ..., n)$ is

$$T = \sum_{jk} a_{jk} \dot{q}_j \dot{q}_k + \sum_j b_j \dot{q}_j + C$$

$$a_{jk} = \frac{1}{2} \sum_i m_i \frac{\partial r_i}{\partial q_j} \cdot \frac{\partial r_i}{\partial q_k}$$

$$b_j = \sum_i m_i \frac{\partial r_i}{\partial q_j} \cdot \frac{\partial r_i}{\partial t}$$

$$C = \frac{1}{2} \sum_i m_i \left(\frac{\partial r_i}{\partial t} \right)^2$$

21. The Lagrangian L is written as a function of q_j and $\dot{q}_j$. If any one coordinate, say q_k, is absent in the expression for the Lagrangian L, then $\dfrac{\partial L}{\partial q_k} = 0$. The equation of motion is

$$\frac{d}{dt} \left(\frac{\partial L}{\partial \dot{q}_k} \right) = 0$$

$$\therefore \frac{\partial L}{\partial \dot{q}_k} = p_k = \text{constant}$$

Here, p_k = linear momentum.

22. The expression for Coriolis force is

$$-2m\omega \times \left(\frac{dr}{dt}\right)_{rot}$$

Here,

m = mass of the particle

ω = angular velocity

23. Consider a rigid body composed of n particles having masses $m_a (a = 1, 2, ..., n)$ and rotating with instantaneous angular velocity ω. Then the expression for the angular momentum due to rotation of the body is

$$L_i = \sum_i I_{ij} \omega_j \qquad (i, j = 1, 2, 3)$$

24. The kinetic energy due to rotation of a body is

$$T = \frac{1}{2}\left[I_{xx}\omega_x^2 + I_{yy}\omega_y^2 + I_{zz}\omega_z^2 + 2I_{xy}\omega_x\omega_y + 2I_{yz}\omega_y\omega_z + 2I_{zx}\omega_z\omega_x\right]$$

25. The equation of the rotational motion of a rigid body in a fixed or inertial frame of reference is

$$\left(\frac{dL}{dt}\right)_{fix} = N$$

Here, N = the torque acting on a rigid body.

26. The torque acting on a rigid body is

$$N = \vec{I} \cdot \dot{\omega} + \omega \times L$$

Here,

L = angular momentum

$\vec{I}$ = moment of inertia

27. The motion of a system from instant t_1 to instant t_2 is such that the line integral is

$$J = \int_{t_1}^{t_2} L \, dt$$

$$L = T - V$$

Hamilton's principle is

$$\delta J = \delta \int_{t_1}^{t_2} L(q_i, \dot{q}_i, t) dt = 0$$

28. In the theory of the canonical transformations, the Lagrangian function is given by

$$L(q, \dot{q}, t) = L'(Q, \dot{Q}, t) + \frac{dF}{dt}(q, Q, t)$$

Here,
Q = new generalized coordinate
F = generating function
L' = new Lagrangian

29. In the new coordinates, Hamiltonian $K = K(Q_K, P_K, t)$. such that $\dot{P}_K = \frac{-\partial K}{\partial Q_K}$ and $\dot{Q}_K = \frac{\partial K}{\partial P_K}$ and the transformations are known as canonical transformations.

Here,
Q_K, P_K = canonical coordinates
Q_K = new position coordinates
P_K = new momentum coordinates

30. The transformed Hamiltonian is

$$K = \sum_i P_i \dot{Q}_i - L'$$

$$= \sum_i P_i \dot{Q}_i - L + \sum_i \left(\frac{\partial F}{\partial q_i} \dot{q}_i + \frac{\partial F_1}{\partial Q_i} \dot{Q}_i \right) + \frac{\partial F_1}{\partial t}$$

$$= H + \frac{\partial F_1}{\partial t}$$

$$p_i = \frac{\partial F_1}{\partial q_i}, \quad P_i = \frac{-\partial F_1}{\partial Q_i}$$

31. $P = \frac{1}{2}(p^2 + q^2)$ and $Q = \tan^{-1}\left(\frac{q}{p}\right)$ are canonical.

32. Consider a generating function of the type $F_k = \sum_k q_k P_k$

then,
$$P_k = \frac{\partial F_2}{\partial q_k}$$
$$= p_k$$

and
$$Q_k = \frac{\partial F_2}{\partial P_k}$$
$$= q_k$$

$K = H$ (canonical transformation)

33. The transformation is canonical if $(pdq - PdQ)$ is an exact differential.

i.e.,
$$pdq - PdQ = d\left(\frac{1}{2}pq\right)$$

Here,
$$P = \frac{1}{2}(p^2 + q^2)$$

$$Q = \tan^{-1}\left(\frac{q}{p}\right)$$

34. The Poisson brackets are defined by the equation

$$[u, v]_{q, p} = \sum_k \left(\frac{\partial u}{\partial q_k}\frac{\partial v}{\partial p_k} - \frac{\partial u}{\partial p_k}\frac{\partial v}{\partial q_k}\right)$$

$$[u, v]_{Q, P} = \sum_k \left(\frac{\partial u}{\partial Q_k}\frac{\partial v}{\partial P_k} - \frac{\partial u}{\partial P_k}\frac{\partial v}{\partial Q_k}\right)$$

35. The Poisson brackets:
    i. are anticommutative, i.e., $[u, v] = -[v, u]$
    ii. are identically zero for a function with itself
    $$[u, u] = 0, [v, v] = 0$$

iii. obey the distributive law
$$[u+v, w] = [u, w] + [v, w]$$

36. The important property of the Poisson bracket is
$$[q_j, p_k] = \delta_{jk}$$
where, δ_{jk} is the Kronecker delta.

37. The canonical equations in terms of the notation of the Poisson bracket are
$$[q_i, H] = \frac{\partial H}{\partial p_i} = \dot{q}_i$$

$$[p_i, H] = \frac{-\partial H}{\partial q_i} = \dot{p}_i$$

38. The total time derivative of a function $u(q, p, t)$ is
$$\frac{du}{dt} = \sum_i \left(\frac{\partial u}{\partial q_i} \frac{\partial H}{\partial p_i} - \frac{\partial u}{\partial p_i} \frac{\partial H}{\partial q_i} \right) + \frac{\partial u}{\partial t}$$

$$\frac{du}{dt} = [u, H] + \frac{\partial u}{\partial t}$$

39. The Jacobi's identity is $[u, [v, w]] + [v, [w, u]] + [w, (u, v]] = 0$

40. The Hamilton–Jacobi equations are
$$H\left(q_1, q_2, \ldots, q_n; \frac{\partial F_2}{\partial q_1}, \frac{\partial F_2}{\partial q_2}, \ldots, \frac{\partial F_2}{\partial q_n}, t\right) + \frac{\partial F_2}{\partial t} = 0$$

where, $F_2 = F_2(q, p, t)$.

The solution of this equation is called Hamilton's principal function and is denoted by S.

i.e., $S = S(q_1, q_2, \ldots, q_n, \alpha_1, \alpha_2, \ldots, \alpha_n, t)$

$\alpha_1, \ldots, \alpha_n$ is additive

$$S = \int L \, dt + \text{constant}$$

41. The Hamilton H of one-dimensional harmonic oscillator is

$$H = \frac{p^2}{2m} + \frac{m\omega^2 q^2}{2} \quad \text{or} \quad \omega = \sqrt{\frac{k}{m}}$$

42. The Hamilton–Jacobi equation can be written as

$$H\left(q_i, \frac{\partial S}{\partial q_i}\right) + \frac{\partial S}{\partial t} = 0$$

The solution $S(q_i, \alpha_i, t) = W(q_i, \alpha_i) - \alpha_1 t$

$$\alpha_1 = \text{constant} = H\left(q_i, \frac{\partial W}{\partial q_i}\right)$$

43. Consider a canonical transformation in which the new momenta are all constants of motion α_i and α_1 is equal to H. Let $W(q, p)$ be the generating function for this transformation. Then the transformation equations are

$$p_i = \frac{\partial W}{\partial q_i}$$

and

$$Q_i = \frac{\partial W}{\partial P_i} = \frac{\partial W}{\partial \alpha_i}$$

$$H(q_i, p_i) = \alpha_1$$

$$p_i = \frac{\partial W}{\partial q_i}$$

$$H\left(q_i, \frac{\partial W}{\partial q_i}\right) = \alpha_1$$

44. The canonical equations for P_i and Q_i are

$$\dot{P}_i = -\frac{\partial K}{\partial Q_i} \quad \text{or} \quad P_i = \alpha_i$$

$$\dot{Q}_i = \frac{\partial K}{\partial P_i} = \frac{\partial K}{\partial \alpha_i} = 1 \quad \text{for } i = 1$$

$$= 0 \quad \text{for } i \neq 1$$

45. Lorentz transformation equations are

$$x' = \frac{x - vt}{\sqrt{1 - \frac{v^2}{c^2}}}$$

$$y' = y$$

$$z' = z$$

$$t' = \frac{t - \frac{vx}{c^2}}{\sqrt{1 - \frac{v^2}{c^2}}}$$

46. The inverse Lorentz transformation can be written as

$$x = \frac{x' + vt'}{\sqrt{1 - \frac{v^2}{c^2}}}$$

$$y = y'$$

$$z = z'$$

$$t = \frac{t' - \frac{vx'}{c^2}}{\sqrt{1 - \frac{v^2}{c^2}}}$$

47. Lorentz–Fitzgerald length contraction equation is

$$L_c = \frac{L}{\sqrt{1 - \frac{v^2}{c^2}}}$$

$$L = L_o \sqrt{1 - \frac{v^2}{c^2}}$$

48. The number of particles scattered (dN) in solid angle ($d\Omega'$) is

$$dN = \sigma(\Omega') \, N d\Omega'$$

Here,

$\sigma(\Omega')$ = proportionality constant (differential cross-section)
N = no. of particles incident per unit area per second on target particles which are at rest

49. In the case of scattering experiments of Rutherford

$$|K| = Z_1 Z_2 e^2$$

where, $Z_1 e$ and $Z_2 e$ are the charges of the incident particles and the target nuclei respectively.

Thus the differential scattering cross-section is given by

$$\sigma(\theta') = \left(\frac{Z_1 Z_2 e^2}{2E}\right) \operatorname{cosec}^4 \frac{\theta'}{2}$$

50. A rigid body with N particles can have $3N$ degrees of freedom, but these are greatly reduced by the constraints, which can be expressed as equations of the form,

$$r_{ij} = c_{ij}$$

Here,

r_{ij} = the distance between the ith and jth particles
c = constant

51. The Euler's equations for the motion of a rigid body with one point fixed are

$$N_x = I_1 \dot{\omega}_x - (I_2 - I_3)\omega_y \omega_z$$

$$N_y = I_2 \dot{\omega}_y - (I_3 - I_1)\omega_z \omega_x$$

$$N_z = I_3 \dot{\omega}_z - (I_1 - I_2)\omega_x \omega_y$$

where,

$\omega_x, \omega_y, \omega_z$ = angular velocity vectors along the principal axes
I_1, I_2, I_3 = principal moments of inertia for the fixed point

52. The equation of Lagrange's function using the Euler's angle is

$$L = T(\dot{\phi}, \dot{\theta}, \dot{\psi}, \phi, \theta, \psi) - V(\phi, \theta, \psi)$$

$$= \frac{1}{2}(I_1 \omega_x^2 + I_2 \omega_y^2 + I_3 \omega_z^2) - V(\phi, \theta, \psi)$$

Here, ϕ, θ, ψ are Euler's angles.

53. The angular velocity components expressed in terms of Euler's angle are

$$\omega_x = \dot{\phi}\sin\theta\sin\psi + \dot{\theta}\cos\psi$$

$$\omega_y = \dot{\phi}\sin\theta\cos\psi - \dot{\theta}\sin\psi$$

$$\omega_z = \dot{\phi}\cos\theta + \dot{\psi}$$

54. The equation for conservation of kinetic energy and angular momentum is

$$\frac{d}{dt}\left(\frac{1}{2}I_1\omega_x^2 + \frac{1}{2}I_2\omega_y^2 + \frac{1}{2}I_3\omega_z^2\right) = 0$$

$$\frac{d}{dt}(iI_1\omega_x + jI_2\omega_y + kI_3\omega_z) = 0$$

55. The kinetic energy of rotation of a rigid body with respect to the principal axes in terms of Eulerian angles is

$$T = \frac{1}{2}I_1\left[\dot{\theta}\cos\psi + \dot{\phi}\sin\theta\sin\psi\right]^2 + \frac{1}{2}I_2\left[-\dot{\theta}\sin\psi + \dot{\phi}\sin\theta\cos\psi\right]^2$$

$$+ \frac{1}{2}I_3\left[\dot{\phi}\cos\theta + \dot{\psi}\right]^2$$

56. If the rigid body is symmetric about one of the principal axes, namely the body z-axis, then the equation of kinetic energy of rotation of a rigid body is

$$T = \frac{1}{2}I_1\left[\dot{\phi}_2\sin^2\theta + \theta\right]^2 + \frac{1}{2}I_3\left[\dot{\phi}\cos\theta + \dot{\psi}\right]^2$$

Here, $I_1 = I_2$

57. A system is said to be in equilibrium if generalized force Q_k acting on the system is zero.

i.e., $$Q_k = \left(\frac{\partial V}{\partial q_k}\right)_0 = 0$$

Here, V = the potential energy.

58. The Lagrangian function of a system of small oscillation is

$$L = \frac{1}{2}\sum_{jk}(T_{jk}\dot{q}_j\dot{q}_k - V_{jk}q_j q_k)$$

59. For Lorentz transformation, the matrix element $a_{\mu\gamma}$ of the transformation between x and x'

$$x'_\mu = \sum_{\gamma=1}^{4} a_{\mu\gamma} x_\gamma$$

60. Lorentz invariant is $(dT)^2 = -\dfrac{1}{c^2}\sum_\mu (dx_\mu)^2$.

61. The relation between the space components of the force K_σ and cartesian components of classical force F is

$$K_i = \frac{F_i}{\sqrt{1-\beta^2}}$$

62. The relativistic definition of classical linear momentum of the particle is

$$p_i = \frac{mv_i}{\sqrt{1-\beta^2}}$$

63. The fourth component of Minkowski force is

$$K_4 = \frac{i}{c}\frac{F\cdot v}{\sqrt{1-\beta^2}}$$

64. The relativistic kinetic energy is

$$T = \frac{mc^2}{\sqrt{1-\beta^2}} + T_o$$

The fourth component of four momentum is $P_\psi = \dfrac{i}{c}T$.

65. The kinetic energy in terms of four momentum is

$$T^2 = p^2c^2 + m^2c^4$$

66. The relativistic mass is

$$m_r = \frac{m_o}{\sqrt{1-\beta^2}}$$

Here, m_o = rest mass of the particle.

67. The relativistic Lagrangian function is
$$L = -mc^2\sqrt{1-\beta^2} - V$$

68. The relativistic Hamiltonian is
$$H = \frac{mc^2}{\sqrt{1-\beta^2}} + V = T + V = E$$

i.e., $H = T + mc^2 + q\phi$

69. The Lagrangian for a single particle in an electromagnetic field is
$$L = -mc^2\sqrt{1-\beta^2} - q\phi + \frac{q}{c}V \cdot A$$

70. The canonical momentum is
$$p_i = mu_i + \frac{q}{c}A_i$$

Here, mu_i is potential on velocity.

71. The co-variant Lagrangian is $\delta I = \delta \int L'(x_\mu, u_\mu, \tau)d\tau = 0$

The co-variant Hamiltonian for a single particle is defined as
$$H' = p_\mu u_\mu - L'$$

The co-variant Hamiltonian in terms of canonical momenta is
$$H' = -\frac{1}{2}mc^2$$

The canonical momentum is $p_4 = \frac{iE}{c}$.

72. The generalized Newton's equation of motion is
$$\frac{d}{dt}(mu_\mu) = K_\mu$$

73. The kinetic energy of the two-coupled oscillator system is
$$T = \frac{1}{2}m_i\dot{x}_1^2 + \frac{1}{2}m_2\dot{x}_2^2$$

74. The Lagrange's two equation of motion for two-coupled oscillator system are
$$m\ddot{x}_2 + m\omega_0^2 x_2 + m\omega_0^2(x_2 - x_1) = 0$$
$$m\ddot{x}_1 + m\omega_0^2 x_1 - m\omega_0^2(x_2 - x_1) = 0$$

75. By the principle of superposition, we can write the general solution for q_j as
$$q_j(t) = \sum_r a_{jr} \exp[i(\omega_r t - \delta_r)]$$
$$q_j(t) = \sum_r a_{jr} \cos[\omega_r t - \delta_r]$$

76. In small oscillation system, near the equilibrium configuration of system, Lagrangian is
$$L = \frac{1}{2}\sum_{j,k}(T_{jk}u_j u_k - V_{jk}u_j u_k)$$
Lagrange's equation of motion is
$$\frac{d}{dt}\left(\frac{\partial L}{\partial u_j}\right) - \frac{\partial L}{\partial u_j} = 0$$
The equation of motion for the coupled system is
$$\sum_k (T_{jK}\ddot{u}_K + V_{jK}u_K) = 0 \quad (j = 1, 2, ..., n)$$
where, u_K = column matrix.

77. Lagrangian in normal coordinate is
$$L = \frac{1}{2}\sum_{i=1}^{n}\dot{n}_i^2 - \frac{1}{2}\sum_{i=1}^{n}\omega l^2 n_i^2$$

78. Lagrange's two equations of motion for two-coupled simple pendulum are
$$ml^2\ddot{\theta}_1 + (mgl + Kl^2)\theta_1 - Kl^2\theta_2 = 0$$
$$ml^2\ddot{\theta}_2 + (mgl + Kl^2)\theta_2 - Kl^2\theta_1 = 0$$

Two frequencies are $\omega_1 = \sqrt{\dfrac{g}{l}}$, $\omega_2 = \sqrt{\dfrac{g}{l} + \dfrac{2K}{m}}$

79. Three normal frequencies of triple pendulum are

$$\omega_1 = \left(\dfrac{g}{l} + \dfrac{2K}{m}\right)^{\frac{1}{2}}, \ \omega_2 = \left(\dfrac{g}{l} + \dfrac{2K}{m}\right)^{\frac{1}{2}}, \ \omega_3 = \left(\dfrac{g}{l} - \dfrac{K}{m}\right)^{\frac{1}{2}}$$

80. For a linear triatomic molecule, i.e., YX_2 molecule, kinetic energy is

$$T = \dfrac{1}{2}m(\dot{q}_1^2 + \dot{q}_3^2) + \dfrac{1}{2}M\dot{q}_2^2$$

$T_{ij} = $ (matrix form)

$$= \begin{pmatrix} m & 0 & 0 \\ 0 & M & 0 \\ 0 & 0 & m \end{pmatrix}$$

Potential energy is

$$V = \dfrac{1}{2}K(q_2 - q_1)^2 - \dfrac{1}{2}K(q_3 - q_2)^2$$

$$V_{ij} = \begin{pmatrix} K & -K & 0 \\ -K & 2K & -K \\ 0 & -K & K \end{pmatrix}$$

Three frequencies are $\omega_1 = 0$, $\omega_2 = \sqrt{\dfrac{K}{m}}$, $\omega_3 = \left\{\dfrac{K}{m}\left(1 + \dfrac{2m}{m}\right)\right\}^{\frac{1}{2}}$

81. The Lagrangian of the small oscillation of particles on string is

$$L = \sum_{j=1}^{n+1}\left[\dfrac{1}{2}m\dot{q}_j^2 - \dfrac{F}{2l}(q_{j-1} - q_j)^2\right]$$

82. A force is said to be central if it acts along the position vector of a particle drawn from the centre of the force. Then it can be represented as $F(r) = \hat{e}_r F(r)$

Here, $\hat{e}_r$ = unit vector along the direction of the position vector r.

83. The equation of motion of a particle subjected to a central force is

$$\mu \ddot{r} = F(r) + \frac{L^2}{\mu r^3}$$

84. The eccentricity is

$$\varepsilon_{\mu_k^2} = \sqrt{1 + 2EL^2}$$

Value of energy	Value of eccentricity	Nature of the orbit
$E > 0$	$\varepsilon > 1$	Hyperbola
$E = 0$	$\varepsilon = 1$	Parabola
$V_{e_{min}} < E < 0$	$0 < \varepsilon < 1$	Ellipse

85. According to Kepler's third law of motion, the period of revolution is

$$T^2 = \frac{4\pi^2}{GM} a^3 = 4\pi^2 \left|\frac{\mu}{K}\right| a^3 \quad \text{(In central force field)}$$

$$T^2 \propto a^3$$

Here,

$$T = \frac{\text{Area}}{\left(\dfrac{dA}{dt}\right)}$$

$$\frac{dA}{dt} = \text{aerial velocity}$$

86. The Kepler's second law of planetary motion in central force field is

$$\text{Aerial velocity} = \frac{dA}{dt} = \frac{1}{2} r^2 \dot{\theta} = \text{constant}$$

87. The differential equation of the orbit is

$$\frac{d^2 x}{d\theta^2} = u - \frac{m}{l^2 u^2} f\left(\frac{1}{u}\right)$$

Here,

$$u = \frac{1}{r}$$

88. The differential cross-section is
$$\sigma(\theta) = \frac{-S}{\sin\theta} \frac{dS}{d\theta}$$
Here, S = ring radius.

89. The Rutherford scattering cross-section is
$$\sigma(\theta) = \frac{1}{4}\left(\frac{ZZ'e^2}{2E}\right)^2 \frac{1}{\sin^4\frac{\theta}{2}}$$
Here,
Ze = charge
$Z'e$ = charge of incident particle

MULTIPLE CHOICE QUESTIONS

1. Relativity predicts a transverse Doppler effect given by

   (a) $\upsilon = \dfrac{\upsilon'}{\sqrt{1-\beta^2}}$
   (b) $\upsilon = \upsilon'\sqrt{1-\beta^2}$
   (c) $\upsilon = \dfrac{\upsilon'}{1+\sqrt{1-\beta^2}}$
   (d) $\upsilon = \upsilon'\left[1+\sqrt{1-\beta^2}\right]$

2. A mechanical system with two bodies in motion is equivalent to a single body with reduced mass m, referred to as the centre of mass. The reduced mass is given in terms of the masses M_1 and M_2 of the two bodies by

   (a) $M = \dfrac{M_1 + M_2}{2}$
   (b) $M = \sqrt{M_1 M_2}$
   (c) $M = \left(\dfrac{M_1}{2} + \dfrac{M_2}{1}\right)$
   (d) $M = \dfrac{M_1 M_2}{M_1 + M_2}$

3. If the acceleration of a particle in the x-direction is given by a_x in a system at rest and if a'_x is the value in another system with relative velocity υ with $\beta = \dfrac{\upsilon}{c}$, a_x and a'_x are related by

   (a) $a_x = a'_x \sqrt{1-\beta^2}$
   (b) $a_x = a'_x(1-\beta^2)$
   (c) $a_x = a'_x(1-\beta^2)^{\frac{3}{2}}$
   (d) $a_x = a'_x(1-\beta^2)^2$

4. Which one of the following particles experiences a Corioli's force?
   (a) a particle at rest with respect to earth at Bhopal
   (b) a particle thrown vertically upwards at Bhopal
   (c) a particle thrown vertically upwards at the north pole
   (d) a particle moving horizontally along the north–south direction at the equator

5. The earth revolves round the sun in an elliptical orbit with sun at one of the foci as shown in figure. The orbital speed of the earth is maximum near the point

   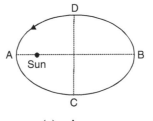

   (a) C (b) D (c) A (d) B

6. A particle is constrained to move along the inner surface of a fixed hemisphere bowl. The number of degrees of freedom of the particle is
   (a) one (b) two (c) three (d) four

7. A body A of mass $2M$ collides with a body B of mass M initially at rest in the laboratory frame. In the centre of mass frame, the two bodies are seen to fly away at right angles to the incident direction. If θ_A and θ_B are the scattering angles for A and B respectively (with respect to the incident direction) in the laboratory frame, then

   (a) $\theta_A < \dfrac{\pi}{2}, \theta_B < \dfrac{\pi}{2}$ (b) $\theta_A < \dfrac{\pi}{2}, \theta_B > \dfrac{\pi}{2}$

   (c) $\theta_A > \dfrac{\pi}{2}, \theta_B < \dfrac{\pi}{2}$ (d) $\theta_A > \dfrac{\pi}{2}, \theta_B > \dfrac{\pi}{2}$

8. A particle has rest mass m_0 and momentum $m_0 c$, where c is the velocity of light. The total energy and velocity of the particles are respectively

   (a) $\sqrt{2} m_0 c^2$ and $\dfrac{c}{2}$ (b) $2 m_0 C^2$ and $\dfrac{c}{\sqrt{2}}$

   (c) $\sqrt{2} m_0 c^2$ and $\dfrac{c}{\sqrt{2}}$ (d) $2 m_0 c^2$ and $\dfrac{c}{2}$

9. In a system of units in which the velocity of light $c = 1$, which of the following is a Lorentz transformation?
   (a) $x' = 4x;\ y' = y;\ z' = z,\ t' = 0.25t$
   (b) $x' = x - 0.5t;\ y' = y;\ z' = z,\ t' = t + x$
   (c) $x' = 1.25x - 0.75t;\ y' = y;\ z' = z,\ t' = 0.75t - 1.25x$
   (d) $x' = 1.25x - 0.75t;\ y' = y;\ z' = z,\ t' = 1.25t - 0.75x$

10. A system is known to be in a state described by the wave function $\psi(\theta, \phi) = \dfrac{1}{\sqrt{30}}(5y_4^0 + y_6^0 - 2y_6^3)$, where $y_l^m(\theta, \phi)$ are spherical harmonics. The probability of finding the system in a state with $m = 0$ is
   (a) zero (b) $\dfrac{2}{15}$ (c) $\dfrac{1}{4}$ (d) $\dfrac{13}{15}$

11. The potential energy of a classical particle moving in one dimension is kx^4 where, k is a constant. If the particle moves from a point x_1 at time t_1 to a point x_2 at time t_2, the actual path followed by the particle is that which makes one of the following integral extremum.
   (a) $\int_{t_1}^{t_2}\left(\dfrac{1}{2}mv^2 + kx^4\right)dt$
   (b) $\int_{t_1}^{t_2}\left(\dfrac{1}{2}mv^2 - kx^4\right)dt$
   (c) $\int_{t_1}^{t_2}\left(\dfrac{1}{2}mv^2 + 4kx^3\right)dt$
   (d) $\int_{t_1}^{t_2}\left(\dfrac{1}{2}mv^2 - 4kx^3\right)dt$

12. In a two body scattering event $A + B \to C + D$, which one of the following is not Lorentz invariant?
   (a) $\dfrac{(P_A + P_B)^2}{C^2}$
   (b) $\dfrac{(P_A - P_B)^2}{C^2}$
   (c) $\dfrac{(P_A - P_C)^2}{C^2}$
   (d) $\dfrac{(P_A - P_D)^2}{C^2}$

13. A particle moves in a circular orbit about the origin under the action of a central force $\vec{F} = \dfrac{-k\vec{r}}{r^3}$. If the potential energy is zero at infinity, the total energy of the particle is
   (a) $\dfrac{-k}{r^2}$ (b) $\dfrac{-k}{2r^2}$ (c) zero (d) $\dfrac{k}{r^2}$

14. In the frame of reference of a rotating turntable, an insect of mass m is moving radially outwards (+ x direction) with a speed v. If the turntable is rotating with a constant angular velocity $w\vec{k}$ in the vertically upward direction, the net pseudo force on the insect, when it is at a distance from the axis of the turntable is given by

    (a) $imw^2r + 2jmwv$
    (b) $imw^2r - 2jmwv$
    (c) $-imw^2r + 2jmwv$
    (d) imw^2r

15. A linear transformation of a generalized coordinate q and the corresponding momentum p to Q and P given by $Q = q + p$, $P = q + \alpha p$ is canonical if the value of the constant α is

    (a) -1 (b) 0 (c) $+1$ (d) $+2$

16. A circle of radius 5 m lies at rest in xy plane in the laboratory. For an observer moving with a uniform velocity v along the y direction, the circle appears to be an ellipse with an equation $\dfrac{x^2}{25} + \dfrac{y^2}{9} = 1$. The speed of the observer in terms of the velocity of light c is

    (a) $\dfrac{9c}{25}$
    (b) $\dfrac{3c}{5}$
    (c) $\dfrac{4c}{5}$
    (d) $\dfrac{16c}{25}$

17. The commutator $[x, p_x^2]$ of the quantum mechanical operator x and p_x^2 is given by

    (a) $2i\hbar$ (b) $2i\hbar p_x$ (c) $-2i\hbar p_x$ (d) $2i\hbar x$

18. The behaviour of the Lorentz force law $\dfrac{d\vec{P}}{dt} = q(\vec{E} + \vec{v} \times \vec{B})$ under transformations of space inversion (P) and time reversal (T) is

    (a) invariant under both P and T
    (b) invariant under P but not under T
    (c) invariant under T but not under P
    (d) invariant under neither P nor T

19. If L is a Lagrangian for a system of n degress of freedom, which Lagrangian yields the same equation of motion?

    (a) $L' = L(q, \dot{q}, t) + f(q)$
    (b) $L' = L(q, \dot{q}, t) + f(\dot{q})$
    (c) $L' = L(q, \dot{q}, t) + \left(\dfrac{df}{dt}\right)(q, t)$
    (d) $L' = L(q, \dot{q}, t)\left(\dfrac{q^2}{2m}\right)$

    where, f is an arbitrary but a differentiable function of its arguments.

20. If the Lagrangian for a system of n degrees of freedom does not contain a given coordinate q_j explicitly, then
    (a) the generalized momentum P_j corresponding to q_j is conserved
    (b) $\dot{q}_j$ is conserved
    (c) q_j is conserved
    (d) momentum conjugate to q_j is zero

21. Two particles of masses 6 kg and 3 kg with position vectors $\vec{r}_1$ and $\vec{r}_2$ respectively are interacting through an interaction potential $u(r)$ where $\vec{r} = \vec{r}_1 - \vec{r}_2$. If $\vec{R}$ denotes the position vector of the centre of mass, then the Lagrangian of the system is
    (a) $L = \dfrac{9}{2}\vec{R}^2 + \vec{r}^2 - u(r)$
    (b) $L = \vec{R}^2 + \vec{r}^2 - u(r)$
    (c) $L = \vec{R}^2 + \dfrac{9}{2}\vec{r}^2 - u(r)$
    (d) $L = 3\vec{R}^2 + 6\vec{r}^2 - u(r)$

22. In elastic collisions the total kinetic energy remains constant. In such a collision with an initially stationary target
    (a) there is transfer of kinetic energy from the target with a corresponding increase in K.E. of the incident particle
    (b) there is transfer of K.E. to the target with a corresponding decrease in the K.E. of the incident particle
    (c) there is no transfer of energy to the target
    (d) the incident particle is brought to rest in the laboratory system

23. The fact that two of the Euler's angles ϕ and ψ are cyclic coordinates in the description of a heavy symmetrical top indicates that
    (a) the torque along the vertical axis is maximum
    (b) the torque along the body z-axis is maximum
    (c) the angular momentum along the vertical and the body z-axis must be constant in time
    (d) the angular momentum along the vertical and the body z-axis is always zero.

24. The general form of a point transformation may be written as
    (a) $Q_i = Q_i(q_1 t)$
    (b) $Q_i = Q_i(q, \dot{q}, t)$
    (c) $Q_i = Q_i(\dot{q}, t)$
    (d) $Q_i = Q_i(q, \dot{q}, \ddot{q}, t)$

25. Consider the Hamiltonian for a simple harmonic oscillator $H = \dfrac{p^2}{2m} + \dfrac{1}{2}Kq^2$. Apply the canonical transformation $P = \dfrac{K}{Q}, q = PQ^2$. Then the resulting Hamiltonian

   (a) has the same form as the original Hamiltonian
   (b) has a different form compared to the original Hamiltonian but has the same value at each point of phase space
   (c) has a different form and different values at each point of phase space
   (d) does not preserve the canonical form of the equations of motion

26. In a system of one degree of freedom, the action variable is defined as

   (a) $J = \oint P dq$
   (b) $J = \oint \dot{q} dP$
   (c) $J = \oint \dot{q} \dot{P} dP$
   (d) $J = \oint \dot{q} \dot{P} dP$

   where the integration is to be carried over a complete period of vibration or of rotation.

27. Denote the matrix of the Lorentz transformation L in the Minkowski space with $L_{\mu\nu}$ as a general element. The orthochronous Lorentz transformation is characterized by

   (a) $L_{\psi j} > 0$
   (b) $L_{\psi j} < 0$
   (c) $L_{\psi\psi} > +1$
   (d) $L_{\psi\psi} < -1$

28. A possible form of the covariant Lagrangian for a relativistic free particle is $\lambda = -mc \sum \sqrt{-X\mu' \times \mu'}$. The corresponding Lagrangian equations are

   (a) $\dfrac{d}{d\tau}(m_{u_v}) = 0$
   (b) $\dfrac{d}{d\tau}\left(m\sqrt{-u_\mu u_v}\right) = 0$
   (c) $\dfrac{d}{d\tau}(mu_v u_\mu) = 0$
   (d) $\dfrac{d}{d\tau}(ml\sqrt{u_v u_\mu}) = 0$

29. The Lagrangian equation of conservative system is

    (a) $\dfrac{d}{dt}\left(\dfrac{\partial L}{\partial \dot{q}_u}\right)+\dfrac{\partial L}{\partial q_u}=0$
    (b) $\dfrac{d}{dt}\left(\dfrac{\partial L}{\partial \dot{q}_k}\right)-\dfrac{\partial L}{\partial q_k}=0$

    (c) $\dfrac{d}{dt}\left(\dfrac{\partial L}{\partial \dot{q}_u}\right)-\dfrac{\partial L}{\partial q_u}=\dot{Q}_k$
    (d) $\dfrac{d}{dt}\left(\dfrac{\partial L}{\partial \dot{q}_u}\right)+\dfrac{\partial L}{\partial q_u}=\dot{Q}_k$

30. The fictitious force acting on a freely falling body of mass 5 kg with reference to a frame moving with a downward positive acceleration of 2 metres/sec² is

    (a) 10 N (b) – 10 N (c) 39 N (d) – 39 N

31. A single particle in phase space is specified by

    (a) 6 coordinates (b) 3 coordinates
    (c) 9 coordinates (d) 12 coordinates

32. The reduced mass of hydrogen atom is almost equal to the

    (a) mass of positronium (b) mass of proton
    (c) half the mass of electron (d) mass of electron

33. $H(P, q)$ and $\overline{H}(P, Q)$ are the Hamiltonian in the coordinates P, q and P, Q respectively. Then, for canonical transformation

    (a) $H(P,q)=\overline{H}(P,Q)$
    (b) $\dot{P}=-\dfrac{\partial \overline{H}}{\partial Q}$

    (c) $\dot{q}=\dfrac{\partial \overline{H}}{\partial P}$
    (d) $H(P,q)=H(P,Q)$

34. The generating function for the transformation $P=\dfrac{1}{Q}, q=PQ^2$ is

    (a) $F=\dfrac{p}{P}$ (b) $F=\dfrac{P}{p}$ (c) $F=\dfrac{q}{Q}$ (d) $F=\dfrac{Q}{q}$

35. A particle of rest mass m_0 is moving with a velocity 0.9c. Its kinetic energy is

    (a) $13m_0c^2$ (b) $0.13m_0c^2$ (c) $1.3m_0c^2$ (d) $130m_0c^2$

36. Any physical law is said to be invariant under Lorentz transformation, if the law can be expressed as a

    (a) covariant three-dimensional equation
    (b) covariant four-dimensional equation

(c) covariant two-dimensional equation

(d) covariant one-dimensional equation

37. If two particles attract each other according to inverse square law of force, then virial theorem implies

(a) $2\bar{T} + \bar{V} = 0$
(b) $\bar{T} + 2\bar{V} = 0$
(c) $2\bar{T} - \bar{V} = 0$
(d) $\bar{T} - 2\bar{V} = 0$

38. The principle of least action states that

(a) $\delta \int_{t_1}^{t_2} L \, dt = 0$
(b) $\delta \int_{t_1}^{t_2} T \, dt = 0$
(c) $\delta \int_{t_1}^{t_2} V \, dt = 0$
(d) $\delta \int_{t_1}^{t_2} (T+V) \, dt = 0$

39. If $\vec{F}_i$ is the force and $\vec{P}_i$ is the momentum then the D'Alembert's principle can be written as

(a) $\sum (\vec{F}_i - \vec{P}_i) \cdot \delta \vec{r}_i = 0$
(b) $\sum_i \vec{F}_i \cdot \dfrac{\delta \vec{r}_i}{\delta q_k} = 1$
(c) $\dfrac{d}{dt} \cdot \dfrac{\delta T}{\partial \dot{q}_k} - \dfrac{\delta T}{\partial q_k} = 0$
(d) $\sum (\vec{F}_i - \vec{P}_i) \cdot \delta \vec{r}_i = 1$

40. The Lagrangian undetermined multiples are used to

(a) eliminate extra virtual work in non-holonomic system
(b) eliminate extra virtual work in holonomic systems
(c) remove constraints in both the holonomic and non-holonomic systems
(d) to include extra virtual work

41. Consider the following statements:

(i) a coordinate that is cyclic is present in the Lagrangian but absent in the Hamiltonian.

(ii) in a rigid body the angular momentum is related to the angular velocity by a quadratic transformation.

Which of the following is correct?

(a) both are true
(b) both are false
(c) only (i) is true
(d) only (ii) is true

42. In the case of elliptical orbits under the influence of a central force, the total energy depends on
    (a) major axis
    (b) minor axis
    (c) both minor and major axes
    (d) neither of the two

43. Fill up the blanks
    (1) For small amplitude oscillations when the extremum of potential energy is a minimum, the equilibrium is _____. (a. stable b. unstable)
    (2) With reference to a two-body central force problem it may be said that the conservation of angular momentum _____ leads to the constancy of a real velocity.
       (a. does b. does not)

44. Kepler's law states that
    (a) the square of the period of revolution of the planet around the sun is proportional to the cube of the semi-major axis.
    (b) the period of revolution is proportional to the area enclosed.
    (c) the square of the period of revolution is proportional to the semi-major axis.
    (d) the square of the period of revolution is inversely proportional to the area enclosed.

45. The Poisson bracket does not obey which of the identities?
    (a) $[u+v, w] = [u, w] + [v, w]$
    (b) $[uv, w] = u[v, w] + [u, w]v$
    (c) $[u, v]_{q_{1p}} = [v, u]_{q_{1p}}$
    (d) $[c_j, P_k] = \delta_{jk}$

46. The proper lifetime of a moon is 2.3×10^{-6} s in a frame in which the velocity of the moon is 0.6 c. The lifetime observed in (s) will be
    (a) 1.38×10^{-6}
    (b) 1.84×10^{-6}
    (c) 2.3×10^{-6}
    (d) 2.88×10^{-6}

47. The transformation $Q = aq + bp, P = cq + ap$ is a canonical transformation if
    (a) $ad - bc = 1$
    (b) $ad - bc = 0$
    (c) $ad + bc = 1$
    (d) $ad + bc = 0$

48. The acceleration of a particle revolving along a circle of constant radius
    (a) increases
    (b) decreases
    (c) remains same
    (d) becomes negative

49. Moment of inertia of a disc of radius R and mass M rotating about an axis is shown in figure.
    (a) $\dfrac{1}{2}MR^2$
    (b) MR^2
    (c) $\dfrac{3}{2}MR^2$
    (d) $2MR^2$

50. The equation of motion of a small particle of mass M' at position x_i is $m\ddot{x} + \gamma\dot{x} - mg = 0$. Assuming the initial speed to be V_i, the terminal speed of the particle will be
    (a) $\dfrac{mg}{\gamma}$
    (b) $\sqrt{v_0 + 2gx}$
    (c) $V_0 + gt$
    (d) $\dfrac{mg}{\gamma^2 t}$

51. Lagrangian of the sun–earth systems is
    (a) $\dfrac{1}{2}m\dot{r}^2 + \dfrac{1}{2}mr^2\dot{\theta}^2 - \dfrac{GMm}{r}$
    (b) $\dfrac{1}{2}m\dot{r}^2 + \dfrac{1}{2}mr^2\dot{\theta}^2 + \dfrac{GMm}{r}$
    (c) $\dfrac{1}{2}m\dot{r}^2 - \dfrac{GMm}{r}$
    (d) $\dfrac{1}{2}mr^2\dot{\theta}^2 + \dfrac{GMm}{r}$

52. Particle A of mass m moving with a velocity v collides head on with particle B at rest and of the same mass m. Assuming elastic collision, which of the following statements is correct for motion after the collision?
    (a) Particle A moves backwards with speed $\dfrac{V}{2}$ and B moves forward with speed $\dfrac{V}{2}$.
    (b) Particle A comes to stop and B moves forward with speed v
    (c) Both of them move forward with speed $\dfrac{V}{2}$
    (d) Both of them move forward with speed $\dfrac{V}{\sqrt{2}}$

53. A particle is moving in $\frac{-1}{r}$ potential. Which of the following statements is incorrect in this case?
    (a) Angular momentum of the particle is always conserved
    (b) Kinetic energy of the particle is always conserved
    (c) The particle always follows a closed path
    (d) Force on the particle is always radial

54. A point mass m is placed at the centre of the hollow sphere of mass M and radius R. The magnitude of the gravitational field inside the sphere but just near the surface is
    (a) $\frac{GM}{R^2}$ (b) $\frac{G(M+m)}{R^2}$ (c) zero   (d) $\frac{Gm}{R^2}$

55. An inertial frame of reference is one which
    (a) is stationary
    (b) moves with uniform velocity
    (c) moves with uniform acceleration
    (d) a and b are correct

56. Lagrange's equations are applicable when the system is
    (a) conservative
    (b) non-conservative
    (c) both conservative and non-conservative
    (d) none of the above

57. The velocity of a rocket of initial mass mv and instantaneous mass m with constant velocity of exhaust gas V.
    (a) is proportional to $\frac{m_o}{m}$
    (b) is inversely proportional to $\frac{m_o}{m}$
    (c) $V = \log\frac{m_o}{M} - \frac{m_o}{M}gt$
    (d) $V \propto \log\frac{m_o}{M}$

58. An observer approaching a source observes the source as if
    (a) the frequency increases
    (b) the frequency decreases

(c) there is no change in the frequency

(d) the frequency is doubled

59. A coordinate is called cyclic when

(a) it is absent in the Lagrangian

(b) it is absent in the Hamiltonian

(c) the conjugate momentum is constant

(d) a, b and c are correct

60. Conservation of four-vector momentum implies conservation of

(a) momentum only

(b) energy only

(c) both (a) and (b)

(d) mass and energy

61. If H is the Hamiltonian and L is the Lagrangian of a system of particles, then the function $(H + L)$ is a function of

(a) the position coordinate only

(b) the velocity only

(c) both position coordinate and velocity

(d) independent of both

62. For the maximum horizontal range, for a missile fired at an angle θ from the horizontal direction, θ is given by : (V is the speed of the missile, g is the acceleration due to gravity)

(a) $\sin\theta = \dfrac{2V^2}{g}$
(b) $45°$
(c) $\sin\theta = \dfrac{V^2}{2g}$
(d) $60°$

63. A body of mass 100 g is orbiting about another body of mass 1 kg due to a central force. What is the reduced mass of the system in the centre of the mass frame?

(a) 0.9 kg

(b) 1.1 kg

(c) 91 g

(d) 110 g

64. A particle is moving in a straight line with a velocity of 10^8 m/sec. Another particle is following it with a velocity of 1.5×10^8 m/sec. What is the relative speed of the second particle with respect to the first one?

(a) 1.67×10^8 m/s

(b) 0.5×10^8 m/s

(c) 2.5×10^8 m/s

(d) 1.0×10^8 m/s

65. It is known that when two bodies move under the influence of a central force between them, the motion is always confined to a plane. This is because
    (a) action and reaction must be equal
    (b) energy must be conserved
    (c) acceleration is always directed towards one point
    (d) angular momentum must be conserved when the force is central

66. The group velocity of a light pulse of average frequency ω, moving in a dispersive medium having refractive index $n(\omega)$ is given by
    (a) $\dfrac{c}{n(\omega)}$
    (b) $cn(\omega)$
    (c) $\dfrac{n(\omega)}{c}$
    (d) $\dfrac{n}{\omega c}$

67. The radii of the circular orbits of the two planets of a star are in the ratio 1 : 3. If the period of the orbit of the nearest planet is 8760 hours, the period of the farther planet is approximately
    (a) 15,154 hours
    (b) 17,600 hours
    (c) 45,464.4 hours
    (d) 26,280 hours

68. Fundamental forces are thought to arise due to exchange of some mediating particles. The range of force λ depends on mass m of mediating particle in the following way
    (a) $\lambda \propto m$
    (b) $\lambda \propto \sqrt{m}$
    (c) $\lambda \propto \dfrac{1}{\sqrt{m}}$
    (d) $\lambda \propto \dfrac{1}{m}$

69. When an alpha particle approaches a nucleus, the trajectory of the scattered alpha particle is
    (a) a hyperbola
    (b) a parabola
    (c) an ellipse
    (d) a circle

70. The integral $\int_{-\infty}^{\infty} \psi^*(i\hbar)\psi\, dx$ represents
    (a) probability density current
    (b) expectation value of energy
    (c) expectation value of positron
    (d) expectation value of momentum

71. If the Lagrangian of a system is $L = \frac{1}{2} e^t (x^2 - \omega^2, x^2)$ the equation of motion is
    (a) $\alpha x + \omega^2 x^2 = 0$
    (b) $\ddot{x} + \alpha \omega^2 - \alpha \omega x = 0$
    (c) $\ddot{x} e^{\alpha t} + \alpha x + \omega^2 x = 0$
    (d) $\ddot{x} + \alpha \dot{x} + \omega^2 x = 0$

72. In relativistic mechanics, the Lagrangian $L(v_i, x_i, t)$ has the properties
    (a) $L = T - V, \dfrac{\partial L}{\partial V_i} = P_i$
    (b) $L \neq T - V, \dfrac{\partial L}{\partial V_i} = P_i$
    (c) $L = T - V, \dfrac{\partial L}{\partial V_i} \neq P_i$
    (d) $L \neq T - V, \dfrac{\partial L}{\partial V_i} \neq P_i$

73. If r_i and r_j are position vectors of two particles in a rigid body, and they are subject to the constraint $(r_i - r_j)^2 - c_{ij}^2 = 0$, where, c_{ij} is a constant, such a constraint is
    (a) holonomic and scleronomous
    (b) holonomic and rheonomous
    (c) non-holonomic and scleronomous
    (d) non-holonomic and rhenomous

74. Two electrons leave a radioactive sample in opposite directions, each with speed 0.5c. With respect to the sample, what is the relativistic speed of one electron with respect to the other
    (a) 0.5c   (b) c   (c) 2c   (d) 0.8c

75. In a generalized coordinate, q_j is cyclic, then in the Lagrangian of the system
    (a) q_j and $\dot{q}_j$ are present
    (b) q_j and $\dot{q}_j$ are necessarily absent
    (c) q_j is absent but $\dot{q}_j$ is not necessarly absent
    (d) q_j is present and $\dot{q}_j$ is absent

76. The motion of two bodies in a central force field is equivalent to that of a single body with a mass equal to
    (a) the mass of the lighter body
    (b) the mass of the heavier body
    (c) the reduced mass
    (d) the average mass

77. A body of mass m_1 collides inelastically with a body of mass m_2 at rest. The ratio of the K.E. is

   (a) $\dfrac{m_1}{m_1 + m_2}$
   (b) $\dfrac{m_2}{m_1 + m_2}$
   (c) $\dfrac{m_1 - m_2}{m_1 + m_2}$
   (d) $\dfrac{\sqrt{m_1 m_2}}{m_1 + m_2}$

78. One of the postulates of the special theory of relativity is that
   (a) there is no preferred inertial frame
   (b) the laws of physics are different for observers in different inertial frames
   (c) the speed of light in vacuum is different in different directions
   (d) the speed of light in vacuum is different in different inertial frames

79. If the time dilation factor for a clock is 2, its speed relative to c is roughly
   (a) 0.63
   (b) 0.75
   (c) 0.87
   (d) 0.99

80. Choose the correct statement concerning the principle of least action.
   (a) The principle of least action deals with the Δ-variation of a physical quantity which is dimensionally angular momentum.
   (b) The motion of a system from time t_i to t_g is in such a way that the line integral of the Lagrangian is an extremum.
   (c) Generalized momentum conjugate to a cyclic coordinate is conserved.
   (d) None of the above.

81. Which is the wrong statement if a wave moves along an ideal stretched string?
   (a) There is no dispersion and the common speed which characterizes the wave is the phase speed of the waves.
   (b) The speed of the wave is independent of the characterisics of the string.
   (c) The speed of a wave is independent of the frequency of the wave.
   (d) The ground speed is the same as the phase speed.

82. If the mach number of a plane is 2, then
    (a) it is flying at a subsonic speed equal to half that of the speed of sound in the atmosphere
    (b) its speed is 1.414 times the speed of the sound in the atmosphere
    (c) it is a supersonic plane whose speed is twice the speed of sound in the atmosphere
    (d) it is a supersonic plane whose speed is four times of the speed of sound in the atmosphere

83. Which of the following is not a solution to the Laplace's equation in two dimensions?
    (a) $\sin(x+iy)$
    (b) $\exp(x)\cos y$
    (c) $10(x^2 - y^2) + 12xy$
    (d) $x^3 - 3x^2 y$

84. The Lagrangian equation for any system containing dissipative forces is
    (a) $\dfrac{d}{dt}\left(\dfrac{\partial L}{\partial \dot{q}_k}\right) - \dfrac{\partial L}{\partial q_k} = 0$
    (b) $\dfrac{d}{dt}\left(\dfrac{\partial L}{\partial \dot{q}_k}\right) - \dfrac{\partial L}{\partial q_k} = Q'_k$
    (c) $\dfrac{d}{dt}\left(\dfrac{\partial L}{\partial \dot{q}_k}\right) - \dfrac{\partial L}{\partial q_k} + \dfrac{\partial R}{\partial \dot{q}_k} = 0$
    (d) $\dfrac{d}{dt}\left(\dfrac{\partial L}{\partial \dot{q}_k}\right) - \dfrac{\partial L}{\partial q_k} + \dfrac{\partial R}{\partial \dot{q}_k} = Q'_k$

85. For conservative systems, the potential energy V does not depend upon
    (a) generalized velocity
    (b) generalized coordinates
    (c) Γ space
    (d) normal coordinates

86. For central force, the orbit always lies in a plane which is perpendicular to the fixed direction of
    (a) linear momentum
    (b) angular momentum
    (c) mass
    (d) radius vector

87. If u is the velocity of a particle and c the velocity of light, its relativistic mass m will exceed its rest mass m_0 by 4% when the ratio $\dfrac{u}{c}$ is
    (a) 0.4
    (b) 0.3
    (c) 0.2
    (d) 0.1

88. The inertial mass of a body is equal to its gravitational mass. This is explained by
    (a) the relativistic law of gravity
    (b) the principle of equivalence
    (c) mass energy relation
    (d) Einstein's gravitational theory

89. The velocity of electrons accelerated by a potential of 1 million volt is
    (a) $\dfrac{2c}{3}$
    (b) $\dfrac{\sqrt{2}c}{3}$
    (c) $\dfrac{3c}{2\sqrt{2}}$
    (d) $\dfrac{2\sqrt{2}c}{3}$

90. The equation of motion of the particle under central force is
    (a) $\dfrac{md^2r}{dt^2} + F(r)\vec{r} = 0$
    (b) $\dfrac{md^2r}{dt^2} - F(r)\vec{r} = 0$
    (c) $\dfrac{md^2r}{dt^2} - F(r)\vec{r} = \text{constant}$
    (d) $\dfrac{md^2r}{dt^2} = 0$

91. The generating function for the transformation $P = \dfrac{1}{Q}; q = pQ^2$ is
    (a) $F = \dfrac{q}{Q}$
    (b) $F = \dfrac{Q}{q}$
    (c) $F = Q \times q$
    (d) $F = \left(\dfrac{q}{Q}\right)^2$

92. A system of particles is in equilibrium only if the total virial work of the actual or applied force is
    (a) constant
    (b) minimum
    (c) zero
    (d) maximum

93. For conservative systems where the coordinate transformation is independent of time, the Hamiltonian function H represents
    (a) K.E. of the system
    (b) P.E. of the system
    (c) total energy of the system
    (d) none of the above

94. The Lagrangian equations for electrical circuit containing finite number of inductances, condensers and resistances is
    (a) $\dfrac{d}{dt}\left(\dfrac{\partial L_{EL}}{\partial \dot{q}_k}\right) - \dfrac{\partial L_{EL}}{\partial q_k} = Q_k$
    (b) $\dfrac{d}{dt}\left(\dfrac{\partial L_{EL}}{\partial \dot{q}_k}\right) - \dfrac{\partial L_{EL}}{\partial q_k} = 0$
    (c) $\dfrac{d}{dt}\left(\dfrac{\partial L_{EL}}{\partial \dot{q}_k}\right) + \dfrac{\partial L_{EL}}{\partial q_k} = Q_k$
    (d) $\dfrac{d}{dt}\left(\dfrac{\partial L_{EL}}{\partial \dot{q}_k}\right) + \dfrac{\partial L_{EL}}{\partial q_k} = 0$

95. Which one of the following is a non-conservative force?
    (a) electrostatic force
    (b) gravitational force
    (c) viscous force
    (d) interatomic force

96. If the Lagrangian L is not an explicit function of time, the Hamiltonian H is
    (a) zero
    (b) constant of motion
    (c) infinity
    (d) variable with motion

97. The reduced mass of positronium is about
    (a) mass of electron
    (b) mass of positron
    (c) mass of hydrogen atom
    (d) one half of the hydrogen atom

98. The moment of inertia is a
    (a) tensor of rank 2
    (b) tensor of rank 1
    (c) not a tensor
    (d) tensor of rank 3

99. If $\{u_e, u_i\}$ is a Lagrange bracket and (u_e, u_j) is a Poisson bracket, then
    (a) $\{u_e, u_i\}(u_e, u_j) = \delta_{ij}$
    (b) $\sum_{l=1}^{2n} \{u_e, u_i\}(u_e, u_j) = \delta_{ij}$
    (c) $\sum_{l=1}^{2n} \{u_e, u_i\}(u_e, u_j) = 0$
    (d) $\{u_e, u_i\}(u_e, u_j) = 0$

100. The force of constraint obeys
    (a) Newton's gravitational law
    (b) Einstein's relativity
    (c) Newton's third law of motion
    (d) friction

101. The energy E, the rest mass m_0 and the momentum p of a relativistic particle are related through the formula
    (a) $E^2 - p^2c^2 = m_0^2 c^2$
    (b) $E^2 - p^2c^2 = m_0^2 c^4$
    (c) $E^2 - pc = m_0^2 c^4$
    (d) $E^2 - pc = m_0^2 c^2$

102. The path followed by a particle in sliding from one point to another in the absence of friction in the shortest time is a
    (a) cycloid   (b) circle
    (c) ellipse   (d) parabola

103. Figure shows three one-dimensional potentials V_1, V_2 and V_3 for a particle. The corresponding frequencies v_1, v_2 and v_3 of small oscillations of the particle about the origin in the 3 cases satisfy

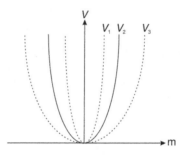

    (a) $v_1 = v_2 = v_3$ (b) $v_1 > v_2 > v_3$
    (c) $v_1 < v_2 < v_3$ (d) $v_1 < v_2 > v_3$

104. The dimensions of a generalized force Q_i are ML^2T^{-2}. The dimensions of the corresponding generalized coordinate q_j and generalized momentum p_j are respectively
    (a) $q_j, \dfrac{\partial L}{\partial \dot{q}_j}$ (b) $\dot{q}_j, \partial \dot{q}_j$ (c) $q_j, \partial L$ (d) $q_j, \partial q_j$

105. A travelling particle is constrained to move in the region $x^2 + y^2 + z^2 < 1$. The number of independent degrees of freedom of the particle is
    (a) 2   (b) 3   (c) 6   (d) 9

106. The SHO under sinusoidal force is given by
    (a) $mx + k\dot{x} = 0$ (b) $m\dot{x} + kx = 0$
    (c) $m\ddot{x} + kx = 0$ (d) $m\ddot{x} + kx = p_0 \sin \omega t$

107. Water flows steadily along a canal of width $2a$. The axis of the canal coincides with the x-axis and its banks are at $y = a$ and $y = -a$. The velocity of the water at any point (x, y) is given by

$\vec{v}(x, y) = k(a^2 - y^2)$ where, k is a positive constant. Which of the following is correct?
(a) the streamlines are straight lines parallel to the x axis
(b) the divergence of the velocity is zero everywhere
(c) the divergence of the velocity is $-2xy$
(d) the flow is irrotational

108. The Lagrangian equations for systems containing dissipative force is

(a) $\dfrac{d}{dt}\left(\dfrac{\partial L}{\partial \dot{q}_k}\right) - \dfrac{\partial L}{\partial q_k} = 0$
(b) $\dfrac{d}{dt}\left(\dfrac{\partial L}{\partial \dot{q}_k}\right) - \dfrac{\partial L}{\partial q_k} = Q'_k$

(c) $\dfrac{d}{dt}\left(\dfrac{\partial L}{\partial \dot{q}_k}\right) - \dfrac{\partial L}{\partial q_k} + \dfrac{\partial k}{\partial \dot{q}_k} = 0$
(d) $\dfrac{d}{dt}\left(\dfrac{\partial L}{\partial \dot{q}_k}\right) - \dfrac{\partial L}{\partial q_k} - \dfrac{\partial k}{\partial \dot{q}_k} = 0$

109. A simple pendulum is taken inside a deep mine. Relative to the period of oscillation on the surface, the time period inside the mine
(a) remains the same
(b) increases
(c) decreases
(d) becomes infinite

110. Consider the sliding of a bead on a circular wire of radius a in the xy plane. The equation of constraint is $x^2 + y^2 = a^2$. This constraint is known as
(a) non-integrable constraint
(b) non-holonomic constraint
(c) holonomic constraint
(d) virtual constraint

111. The reduced mass of HCl molecule in amu is
(a) 97 amu
(b) 0.97 amu
(c) 9.7 amu
(d) 0.097 amu

112. The Lagrangian equation when the Lagrangian function has the form $L = T-U$ is
(a) $\dot{q}_k = 0$
(b) $\ddot{q}_k = 0$
(c) $q_k = 0$
(d) $q_k \neq 0$

113. The motion of a simple pendulum undergoing large oscillations can be described as
    (a) holonomic, non-conservative
    (b) harmonic, conservative
    (c) anharmonic, non-conservative
    (d) anharmonic conservative

114. Coordinates that do not appear explicitly in the Lagrangian of a system are said to be
    (a) cyclic   (b) cartesian
    (c) cylindrical   (d) normal

115. For a conservative system the potential energy V does not depend upon
    (a) generalized velocity   (b) acceleration
    (c) virtual work   (d) force

116. The gravitational field due to an infinitely long, thin, straight rod with uniform mass distribution varies with distance from the rod as
    (a) $\dfrac{1}{r^2}$ (b) $\dfrac{1}{r}$
    (c) $\dfrac{1}{\sqrt{r}}$ (d) independent of r

117. If the particles attract each other, according to inverse square law of force
    (a) $T + V = 0$ (b) $T - V = 0$
    (c) $2T + V = 0$ (d) $2T - V = 0$

118. For a central force, the orbit always lies in a plane which is perpendicular to the fixed direction of
    (a) linear momentum   (b) mass
    (c) radius vector   (d) angular momentum

119. When a planet moves around the sun, its
    (a) aerial velocity is constant
    (b) aerial velocity depends on its position
    (c) linear velocity is constant
    (d) angular velocity is constant

120. A particle of mass m under the action of a force describes an orbit $r = r_0 e^\theta$. The form of the force function that leads to this spiral orbit is

(a) $-\dfrac{2l^2}{\mu r^3}$ (b) $-\dfrac{2l}{\mu r}$ (c) $\dfrac{2l^2}{\mu r^3}$ (d) $\dfrac{2l}{\mu r}$

121. A cylindrical object (length L) of density d_c floats in a liquid of density d_e. The natural frequency ω_0 of oscillation when the cylinder is depressed by an external force and then released is

(a) $\dfrac{gd_e}{Ld_c}$ (b) $\dfrac{g}{Ld_c}$ (c) $\dfrac{d_e}{Ld_c}$ (d) $\dfrac{\sqrt{gd_e}}{Ld_c}$

122. When the two frequencies are identical in small oscillations, the system is known as

(a) resonance (b) oscillator
(c) degenerate (d) linear

123. Euler's equation predicts that the shortest distance between two fixed points in a plane is a

(a) circle (b) curve
(c) plane (d) straight line

124. The generalized function for the transformation $P = \dfrac{1}{Q}$, $q = PQ^2$ is

(a) $F = \left(\dfrac{q}{Q}\right)^2$ (b) $F = Q \times q$

(c) $F = \dfrac{Q}{q}$ (d) $F = \dfrac{q}{Q}$

125. The kinetic energy of a relativistic particle is twice its rest mass energy, the speed of the particle is

(a) $\dfrac{2\sqrt{2}}{3c}$ (b) $\left(\dfrac{2\sqrt{2}}{3}\right)c$ (c) $\left(\dfrac{1}{3}\right)c$ (d) $\left(\dfrac{1}{\sqrt{3}}\right)c$

126. Two electrons move towards each other, the speed of each being 0.9c in a Gallilean frame of reference. Their speed relativistic to each other is

(a) 0.4c   (b) 0.995c   (c) – 0.995c   (d) 1.2c

127. q_1 and q_2 are generalized coordinates and P_1 and P_2 are the corresponding generalized momenta. The Poisson bracket (X, Y) of $X = q_1^2 + q_2^2$ and $Y = 2P_1 + P_2$ is

   (a) $(q_1^2 + q_2^2)P_1$
   (b) $3(q_1^2 + q_2^2)$
   (c) $4q_1 + 2q_2$
   (d) 0

128. The number of independent coordinates required to describe the motion of a rigid body is

   (a) 2   (b) 6   (c) 3   (d) 1

129. The escape velocity from the surface of a spherical planet of mass M is given by $\dfrac{\sqrt{GM}}{2R}$, the radius of the planet is

   (a) $\dfrac{R}{2}$ (b) R (c) $2R$ (d) $4R$

130. Let $\vec{E}(\vec{r}, t)$ and $\vec{B}(\vec{r}, t)$ denote electric and magnetic fields obeying Maxwell's equations. Then under a Lorentz transformation from one inertial frame to another

   (a) $\varepsilon_o E^2 + \dfrac{B^2}{\mu_o}$ is invariant
   (b) $\varepsilon_o E^2 - \dfrac{B^2}{\mu_o}$ is invariant
   (c) $(\vec{E} \cdot \vec{B})^2$ is invariant
   (d) $(\vec{E} \times \vec{B})^2$ is invariant

131. The frequency of a man's voice is 165 Hz and the wavelength of his sound wave is 2 m. If the wavelength of the sound wave produced by a child's voice is 3 m, the frequency of the child's voice is

   (a) 248 Hz   (b) 165 Hz   (c) 110 Hz   (d) 250 Hz

132. Which of the following properties of fluids can serve to distinguish between a gas and a liquid?

   (a) immiscibility
   (b) malleability
   (c) viscosity
   (d) compressibility

133. Which of the following is true when the pendulum of a clock reaches the highest point of its arc?

   (a) the net force acting on the system is zero
   (b) the kinetic energy is maximum

(c) the potential energy is maximum

(d) the frequency is zero

134. A system is executing damped vibrations and is further subjected to an external periodic force differing in frequency. Then the total system

   (a) will be in resonance

   (b) will not be in resonance

   (c) will be underdamped

   (d) will be overdamped

135. End of glass tube becomes round on heating due to

   (a) surface tension   (b) viscosity

   (c) gravity   (d) friction

136. A plane travelling at Mach number 2 is moving at

   (a) half the speed of sound

   (b) the speed of sound

   (c) twice the speed of sound

   (d) four times the speed of sound

137. A rubber balloon inflated with hydrogen gas is released at a height of 2 metres from the surface of the moon. The balloon will

   (a) come down

   (b) initially rise and then come down

   (c) rise upwards

   (d) continue to stay at a height of 2 metres

138. The statement that space possesses translation symmetry leads to

   (a) conservation of angular momentum

   (b) conservation of linear momentum

   (c) conservation of energy

   (d) conservation of parity

139. The statement that space possesses time displacement symmetry leads to

   (a) conservation of parity

   (b) conservation of angular momentum

   (c) conservation of linear momentum

   (d) conservation of energy

140. For a particle moving in a central force field
    (a) aerial velocity is constant
    (b) linear momentum is constant
    (c) linear velocity is constant
    (d) none of the above

141. For a conservative system the Lagrangian equation of motion in terms of generalized coordinates q and momentum p is
    (a) $\dfrac{\partial L}{\partial q_j} - \dfrac{\partial L}{\partial p_j} = 0$
    (b) $\dfrac{\partial L}{\partial \dot{q}_j} - \dfrac{\partial}{\partial t}\left(\dfrac{\partial L}{\partial q_j}\right) = 0$
    (c) $\dfrac{\partial}{\partial t}\left(\dfrac{\partial L}{\partial q_j}\right) - \dfrac{\partial}{\partial t}\left(\dfrac{\partial L}{\partial \dot{q}_j}\right)$
    (d) $\dfrac{\partial}{\partial t}\left(\dfrac{\partial L}{\partial \dot{q}_j}\right) - \left(\dfrac{\partial L}{\partial q_j}\right) = 0$

142. X, Y, Z is a stationary frame and X', Y', Z' is a rotating frame rotating with an angular velocity $\bar{\omega}$ and Z, Z' axes coinciding with each other. A particle is rotating with an angular velocity $\bar{\omega}$ in X, Y, Z frame in XY plane. Its motion in X', Y', Z' frame can be represented as
    (a) rotating with $-\bar{\omega}$
    (b) rotating with $2\bar{\omega}$
    (c) stationary
    (d) rotating with $-2\bar{\omega}$

143. To specify the configuration of a rigid body containing N particles, the following independent generalized coordinates are required
    (a) $3N$
    (b) $3N - 6$
    (c) 6
    (d) $\dfrac{N}{3}$

144. The average kinetic energy of a system of non-interacting particles/degree of freedom is
    (a) $\dfrac{2kT}{2}$
    (b) $\dfrac{kT}{2}$
    (c) $\dfrac{2kT}{3}$
    (d) $\dfrac{3nkT}{2}$

145. Relativistic motion of a particle in an attractive inverse square law of force is
    (a) ellipse
    (b) processing ellipse
    (c) parabola
    (d) circle

146. The momentum of a photon with energy E is
    (a) $\dfrac{E}{h} = p$
    (b) $\dfrac{E}{c} = p$
    (c) $\dfrac{h}{E} = p$
    (d) $\dfrac{c}{E} = p$

147. The Galilean transformation equation for acceleration of a particle in two frames of references S' and S moving with a relative velocity v is given by

(a) $a'_x = \dfrac{ax}{\sqrt{1-\dfrac{v^2}{c^2}}}$

(b) $a'_x = ax\sqrt{1-\dfrac{v^2}{c^2}}$

(c) $a'_x = ax\sqrt{1-\dfrac{v}{c}}$

(d) $a'_x = ax$

148. In Minkowsky space Lorentz transformations are simply the
   (a) orthogonal transformation
   (b) transformation from cartesian to polar
   (c) transformation from cartesian to generalized
   (d) cartesian to spherical polar

149. The most general displacement of a rigid body is
   (a) translational
   (b) rotational
   (c) translational and rotational
   (d) translational and vibrational

150. The angular speed of minutes hand of a wrist watch in radians per second is given by

(a) $\dfrac{\pi}{60 \times 60}$ (b) $\dfrac{\pi}{60 \times 30}$ (c) $\dfrac{\pi}{60}$ (d) $\dfrac{2\pi}{60}$

ANSWERS

1. (d)	2. (d)	3. (c)	4. (b)	5. (c)
6. (b)	7. (a)	8. (c)	9. (d)	10. (d)
11. (b)	12. (a)	13. (c)	14. (d)	15. (c)
16. (c)	17. (b)	18. (b)	19. (c)	20. (a)
21. (a)	22. (a)	23. (c)	24. (a)	25. (b)
26. (a)	27. (c)	28. (d)	29. (b)	30. (c)
31. (a)	32. (d)	33. (a)	34. (c)	35. (c)

36. (b)	37. (a)	38. (b)	39. (a)	40. (c)
41. (d)	42. (a)	43. i (a), ii (b)	44. (a)	45. (c)
46. (d)	47. (a)	48. (b)	49. (a)	50. (a)
51. (a)	52. (a)	53. (d)	54. (b)	55. (b)
56. (c)	57. (d)	58. (c)	59. (d)	60. (c)
61. (c)	62. (b)	63. (c)	64. (c)	65. (d)
66. (a)	67. (d)	68. (d)	69. (b)	70. (b)
71. (d)	72. (a)	73. (a)	74. (d)	75. (c)
76. (c)	77. (a)	78. (b)	79. (c)	80. (d)
81. (c)	82. (b)	83. (b)	84. (c)	85. (b)
86. (b)	87. (b)	88. (b)	89. (d)	90. (b)
91. (a)	92. (c)	93. (c)	94. (a)	95. (c)
96. (b)	97. (d)	98. (a)	99. (b)	100. (c)
101. (b)	102. (a)	103. (c)	104. (a)	105. (b)
106. (d)	107. (c)	108. (c)	109. (b)	110. (c)
111. (b)	112. (b)	113. (c)	114. (a)	115. (a)
116. (a)	117. (c)	118. (d)	119. (a)	120. (a)
121. (d)	122. (c)	123. (d)	124. (d)	125. (b)
126. (c)	127. (c)	128. (c)	129. (d)	130. (c)
131. (c)	132. (d)	133. (c)	134. (b)	135. (a)
136. (c)	137. (b)	138. (b)	139. (d)	140. (a)
141. (d)	142. (c)	143. (c)	144. (b)	145. (d)
146. (a)	147. (d)	148. (a)	149. (c)	150. (b)

2

STATISTICAL PHYSICS

FORMULAE

1. The entropy of the system is given by
$$S = K \log_e W$$
Here,

   K = Boltzmann's constant

   W = number of states accessible to the system

2. Mean value of an ensemble is
$$\left\langle q^2 \right\rangle_e = \frac{1}{N} \sum_i (q_i)^2$$
Here,

   N = No. of identical systems

   q_i = coordinate of the particle in the ith spin

   e = mean taken over the entire system

3. The enthalpy of any system is
$$H = U + pv$$
Here,

   U = total internal energy of the system

   p, v = state variables

4. Helmholtz function or Helmholtz free energy is
$$F = U - Ts$$
Here,
H = enthalpy of the system
Ts = latent energy or bound energy of the system

5. Gibb's potential or Gibb's free energy
$$G = H - Ts$$
Here,
H = enthalpy of the system
Ts = latent energy or bound energy of the system

6. The chemical potential which is defined as the rate of change of free energy per mole at constant volume and temperature is
$$\mu = \left(\frac{\partial F}{\partial N_i}\right)_{T,V,N_j}$$

7. The occupation number for particles obeying quantum statistics is
$$\overline{n} = \frac{1}{e^{\beta(\varepsilon-\mu)} + K}$$
(K = +1 for fermions and −1 for bosons)
Here,
ε = energy per particle in a particular state
μ = chemical potential

8. Partition function is
$$Z = \sum_i \exp(\beta \varepsilon_i)$$
$$\beta = (k_B T)^{-1}$$

9. The single-particle partition function is
$$Z_s = \left(\frac{m}{2\beta\pi\hbar^2}\right)^{\frac{3}{2}} V$$
Here,
V = volume

$\beta = (k_B \pi)^{-1}$

m = total mass of the molecule

10. Partition function for a diatomic molecule is

$$Z_D = \sum_{\text{(states)}} e^{-\beta \varepsilon_i}$$

$$\varepsilon_i = \varepsilon_{tr} + \varepsilon_{rot} + \varepsilon_{vib} + \varepsilon_e + \varepsilon_n$$

Here,

ε_{tr} = translational energy of the centre of mass of the molecule

ε_{rot} = rotational energy associated with the rotation of constituent atoms in the molecule about the centre of mass

ε_{vib} = vibrational energy of the two atoms along the line joining them

ε_e = energy of the atomic electrons

ε_n = energy of the atomic nucleus

(a) Translational partition function is

$$Z_{tr} = \left(\frac{m}{2\beta\pi\hbar^2}\right)^{\frac{3}{2}} V$$

(b) Rotational partition function is

$$Z_{rot} = \int_0^\infty (2j+1)\exp[-j(j+1)\theta_{rot}/T]\, dj$$

(c) Vibrational partition function is

$$Z_{vib} = e^{-\frac{\theta_{vib}}{2\pi}} \left[1 - e^{-\frac{\theta_{vib}}{T}}\right]^{-1}$$

$$\theta_{vib} = \left(\frac{\hbar\omega}{k_\beta}\right)$$

(d) Electronic partition function is

$$Z_e = \sum_{\text{levels}} g_e e^{-\beta \varepsilon_e}$$

(e) Nuclear partition function
$$Z_n = g_n$$
where, g_n = degeneracy due to nuclear spin.
Total partition function is
$$Z_D = \left(\frac{m}{2\pi\beta t^2}\right)^{\frac{3}{2}} V Z_{int}$$

11. Relation between partition function and entropy is
$$S = KN \log Z + \frac{3}{2}NK$$
Here,
 N = total number of molecules
 K = Boltzmann constant
 Z = partition function
 S = entropy

12. Relation between F and Z is
$$F = -NKT \log Z$$
Here,
 F = Helmholtz free energy
 N = total number of molecules
 T = temperature
 K = Boltzmann constant
 Z = partition function

13. Relation between Z and E is
$$E = NKT^2 \left[\frac{\partial}{\partial t}(\log Z)\right]_V$$
Here, E = total energy.

14. Relation between Z and H is
$$H = NKT^2 \left[\frac{\partial}{\partial t}(\log Z)\right]_V + RT$$

Here,
- R = gas constant
- Z = partition function
- H = enthalpy

15. Relation between G and Z is

$$G = RT - NKT \log Z$$

Here,
- G = Gibb's free energy
- R = gas constant

16. Bose–Einstein statistics is

$$n_i = \frac{g_i}{e^{\alpha + \beta \varepsilon_i} - 1}$$

Here, g_i = degeneracy of the ith level.

17. Fermi–Dirac statistics is

$$n_i = \frac{g_i}{e^{(\alpha + \beta \varepsilon_i)} + 1}$$

18. Maxwell–Boltzmann statistics is

$$n_i = \frac{g_i}{e^{\alpha + \beta \varepsilon_i}}$$

19. Specific heat at constant volume is

$$C_v = \left(\frac{\partial E}{\partial T}\right)_v$$

20. The Ising model energy expression is

$$E = -J \sum_{nn} S_i S_j$$

Here, S = spin.

21. Relation between the specific heat capacity at constant volume C_v and pressure C_p is

$$C_p - C_v = VT \frac{\alpha^2}{k}$$

Here,

α = volume coefficient of expansion

$$\alpha = \frac{1}{V}\left(\frac{\partial V}{\partial T}\right)_P$$

K = isothermal compressibility

$$K = -\frac{1}{V}\left(\frac{\partial V}{\partial P}\right)_T$$

22. The canonical distribution is

$$P_r \propto e^{-\beta E_r}$$

Here,

P_r = probability of finding a particle in state r
E_r = energy of a particle in state r
β = temperature

23. The grand canonical distribution is

$$P_r \propto e^{-\beta E_r - \alpha N_r}$$

Here,

β = temperature; $(\beta = (KT)^{-1})$

24. The equipartition function of classical statistical mechanics is

$$\overline{\varepsilon_i} = \frac{1}{2}kT$$

25. The Clausius–Clapeyron equation is given by

$$\frac{dp}{dT} = \frac{\Delta S}{\Delta v}$$

Here,

$\Delta S = S_2 - S_1$ = change in entropy
$\Delta v = v_2 - v_1$ = change in volume

26. Gibb's–Duhem relation is

$$SdT - Vdp + \sum_i N_i d\mu_i = 0$$

Here,

μ = chemical potential
N_i = number of molecules of type i
V = volume
S = entropy of the system

27. The general condition for chemical equilibrium is

$$\sum_{i=1}^{m} b_i \mu_i = 0$$

Here, μ_i = chemical potential.

28. Planck distribution is

$$\overline{n}_S = \frac{1}{e^{\beta E_S} - 1}$$

29. The general formula of the fluctuation dissipation theorem is

$$\alpha_{ij} = \frac{1}{k} \int_{-\infty}^{0} ds\, K_{ik}(s)$$

Here, α_{ij} = friction coefficient.

30. Liouville's theorem is given by

$$\frac{dp}{dt} = \frac{\partial p}{\partial t} + \sum_i \left(\frac{\partial p}{\partial q_i} \dot{q}_i + \frac{\partial p}{\partial p_i} \dot{p}_i \right) = 0$$

where, $\frac{dp}{dt}$ measures the rate of change of p if one moved along in phase space with the point representing a system.

MULTIPLE CHOICE QUESTIONS

1. Planck's law of energy distribution of black body radiation agrees with the Rayleigh–Jeans law
   (a) at all wavelengths
   (b) at short wavelengths only
   (c) at long wavelengths only
   (d) only at the maximum of the energy distribution curve

2. In Debey's model of vibrations of solid the minimum wavelength is equal to

   (a) a (b) a^3 (c) $2a$ (d) $\sqrt{a}$

   where, a is the interatomic distance.

3. The Maxwell–Boltzmann distribution law is represented by

   (a) $n_v = C \exp\left(\dfrac{-mv^2}{2kT}\right)$ (b) $n_v = C^2 \exp\left(\dfrac{-mv^2}{2kT}\right)$

   (c) $n_v = C^2 \exp\left(\dfrac{-mv^2}{kT}\right)$ (d) $n_v = C \exp\left(\dfrac{-mv}{kT}\right)$

4. At low temperature, the specific heat of hydrogen is due to
   (a) rotational states
   (b) translation
   (c) vibrational levels
   (d) electronic levels

5. According to Fermi–Dirac statistics, at the Fermi energy, the occupation index is
   (a) zero
   (b) almost zero
   (c) half
   (d) unity

6. Einstein's theory of specific heat predicts values lower than the experimental values at
   (a) all temperatures
   (b) very low tempertures
   (c) very high temperatures
   (d) intermediate temperatures

7. Bose–Einstein condensation temperature T_B refers to the temperature below which
   (a) an assembly of Bose gas condense to the liquid state.
   (b) there is an appreciable occupation of the ground state in an electron system.
   (c) there is a significantly large occupancy of the ground state in a system of bosons.
   (d) the bosons essentially behave like fermions.

8. An engine absorbs heat at a temperature of 1000 K and rejects heat of 600 K. If the engine operates at maximum possible efficiency, the amount of work performed by the engine, for 2000 J heat input is

(a) 1600 J   (b) 1200 J   (c) 800 J   (d) 400 J

9. A system has N distinguishable particles. Each particle can occupy one of the two non-degenerate states with an energy difference of 0.1 eV. If the system is in thermal equilibrium at room temperature, the approximate fraction of particles in the higher energy state is

(a) exp (−10)   (b) exp (−4)   (c) exp (−2)   (d) zero

10. In an ideal gas adiabatic expansion, if the volume of the gas doubles from V_0 to $2V_0$, what happens to the temperature?

(a) increases to $1.63T_0$ (b) falls to $0.63T_0$
(c) remains the same   (d) falls to $0.36T_0$

11. For a system at a constant temperature and pressure the equilibrium state is described by

(a) minimum of Helmoltz free energy
(b) minimum of Gibbs potential
(c) minimum of entropy
(d) minimum of volume

12. A 10 g specimen of copper at temperature 350 K is placed in thermal contact with a second 10 g specimen of copper at 290 K. The quantity of energy transferred when the two specimens are placed in contact is (specific heat of Cu = 3.89×10^6 ergs gm^{-1} deg^{-1} and one calorie = 4.184×10^7 ergs)

(a) 3.89×10^7 ergs   (b) 1.17×10^8 ergs
(c) 4.184×10^8 ergs   (d) 4.184×10^7 ergs

13. The average energy for a photon gas at temperature T when the energy levels are given by $E_n = n\hbar\omega$ $(n = 0, 1, 2,...)$ is

(a) $\dfrac{\hbar\omega}{(e^{\hbar\omega\beta} - 1)}$ (b) $\dfrac{\hbar\omega}{(1 - e^{-\hbar\omega\beta})}$

(c) $\dfrac{1}{(1 - e^{-\hbar\omega\beta})}$ (d) $\dfrac{\hbar\omega}{(e^{\hbar\omega\beta} - e^{-\hbar\omega\beta})}$

14. The total entropy σ of a combined system of two closed systems with entropies σ_1 and σ_2 is
    (a) $\sigma = \sigma_1 + \sigma_2$
    (b) $\sigma = \sigma_1 \cdot \sigma_2$
    (c) $\sigma = |\sigma_1 - \sigma_2|$
    (d) $\sigma = \log(\sigma_1 - \sigma_2)$

15. A thermally insulated ideal gas ($dQ = 0$) is compressed quasi-statically from an initial macrostate of volume V_0 and pressure P_0 to a final macrostate of volume V_1 and pressure P_1. Then,
    (a) $\left(\dfrac{C_V}{R}\right)(P_1 V_1 - P_0 V_0)$
    (b) $\left(\dfrac{C_P}{R}\right)(P_1 V_1 - P_0 V_0)$
    (c) $\left(\dfrac{C_V}{C_P}\right)(P_1 V_1 - P_0 V_0)$
    (d) $\dfrac{1}{(P_1 V_1 - P_0 V_0)}$

16. Use of Fermi gas model for electrons in a metal of density 0.971 g/cm^2 and molar mass 22.99 g/mol to obtain a Fermi momentum of
    (a) 0.91 A^{-1}
    (b) 0.22 A^{-1}
    (c) 0.11 A^{-1}
    (d) 0.48 A^{-1}

17. For temperatures $T < T_0$ (T_0 is the Bose–Einstein temperature) the number of atoms in the excited orbitals varies as
    (a) $T^{\frac{1}{2}}$
    (b) T
    (c) $T^{\frac{3}{2}}$
    (d) T^2

18. At absolute zero the pressure of an ideal Fermi gas is
    (a) 0
    (b) $\dfrac{2 E_F}{5 V}$
    (c) $\dfrac{E_F}{V}$
    (d) $\sqrt{\dfrac{E_F}{V}}$

19. Use Boltzmann factor to study the thermodynamics of N independent particles of a spin system in a magnetic field where the energies are $E\pm = \mp\mu_0 B$. Then the total energy at inverse temperature $\beta = \dfrac{1}{kT}$ is
    (a) $-N\mu_0 B \tanh(\beta\mu_0 B)$
    (b) $-N\mu_0 B \cosh(\beta\mu_0 B)$
    (c) $-N\mu_0 B \coth(\beta\mu_0 B)$
    (d) $-N\mu_0 B \sinh(\beta\mu_0 B)$

20. The interaction between two gases A and A' is thermal interaction. The walls which allow thermal interaction are called
    (a) adiabatic walls
    (b) isothermal walls
    (c) diathermic walls
    (d) isobaric walls

21. In thermodynamics, temperature can be expressed as
    (a) $T = \left(\dfrac{\partial u}{\partial S}\right)_V$
    (b) $T = \left(\dfrac{\partial S}{\partial u}\right)_V$
    (c) $T = \Delta Q$
    (d) $T = \dfrac{\Delta S}{\Delta Q}$

22. The average energy of a bosonic harmonic oscillator in thermal equilibrium with a heat bath at temperature T is
    (a) $\dfrac{1}{2}\hbar\omega$
    (b) $\dfrac{\hbar\omega}{e^{\frac{\hbar\omega}{kT}} - 1}$
    (c) $\dfrac{1}{e^{\frac{\hbar\omega}{kT}} - 1}$
    (d) $\dfrac{\hbar\omega}{e^{\frac{\hbar\omega}{kT}} + 1}$

23. The partition function Z_{ij} of two independent systems i and j is given by
    (a) $Z_{ij} = Z_i \times Z_j$
    (b) $Z_{ij} = Z_i + Z_j$
    (c) $Z_{ij} = Z_i - Z_j$
    (d) $Z_{ij} = \dfrac{Z_i}{Z_j}$

24. The volume element of phase space $\Delta \tau$ of a single spinless particle in 3-dimension is given by
    (a) $\dfrac{4\pi P^2}{h^3}$
    (b) $\dfrac{4\pi P^2 dP}{h^3}$
    (c) $\dfrac{4\pi P^2 V}{h^3}$
    (d) $\dfrac{4\pi P^2 dPV}{h^3}$

25. The number of ways in which N objects can be distributed in K boxes with n_i ($i \leq i \leq k$) objects in ith box is given by
    (a) $N!$
    (b) $n_1! n_2! n_3! \cdots n_k!$
    (c) $\dfrac{N!}{n_1! n_2! n_3! \cdots n_k!}$
    (d) $\dfrac{n_1! n_2! n_3! \cdots n_k!}{N!}$

26. The pressure of element gas at 0 K is
   (a) 3.8×10^5 atm.
   (b) 3.8 atm.
   (c) 38 atm.
   (d) $3.8\ 10^{-5}$ atm.

27. At room temperatures, the molar heat capacity of all solids is nearly
   (a) R
   (b) $2R$
   (c) $3R$
   (d) $4R$

28. Transport properties of liquid Helium II in normal state are similar to those of
   (a) solid
   (b) gas
   (c) liquid
   (d) vapour

29. The average number of photons in an enclosure of 22.4 litres at 273 K is
   (a) 9.17
   (b) 917
   (c) 9.17×10^{10}
   (d) 9.17×10^{12}

30. The zeroth law of thermodynamics enables us to give a precise meaning to
   (a) pressure
   (b) temperatures
   (c) entropy
   (d) free energy

31. If a system is such that it neither shows a tendency to undergo a spontaneous change in its internal structure (such as during a chemical reaction) nor allows transfer of matter from one portion of it to another (such as in diffusion), it is said to be in
   (a) chemical equilibrium
   (b) thermal equilibrium
   (c) mechanical equilibrium
   (d) statical equilibrium

32. The vapour pressure (in millimetres of mercury) of solid ammonia is given by $I_n P = 23.03 - \dfrac{3754}{T}$ and that of the liquid ammonia by $I_n P = 19.49 - \dfrac{3063}{T}$. The temperature of the triple point is
   (a) 10 K
   (b) 100 K
   (c) 195 K
   (d) 300 K

33. For a reversible isothermal and isobaric process
   (a) $\Delta E = 0$
   (b) $\Delta S = 0$
   (c) $\Delta F = 0$
   (d) $\Delta G = 0$

   where E, S, F and G stand for the internal energy, entropy, Helmholtz potential and Gibbs potential respectively.

34. Consider a system consisting of two bosons each of which can be in any of the two quantum of respective energies 0 and t. The partition function of the system is given by

(a) $\left[1+e^{-\frac{\varepsilon}{kT}}\right]^2$

(b) $1+e^{-\frac{\varepsilon}{kT}}+e^{-\frac{2\varepsilon}{kT}}$

(c) $e^{-\frac{\varepsilon}{kT}}$

(d) none of the above

35. In the case of a free gas of Fermi particles, the quantum exchange effects lead to the occurence of
   (a) an additional effective attraction between the particles at the low temperature regime
   (b) an additional effective attraction between the particles at the higher temperature regime
   (c) an additional effective repulsion between the particles
   (d) none of the above

36. In Fermions, the spin of the nucleus is
   (a) an integral multiple of $\frac{1}{2}$
   (b) zero
   (c) even integral multiple of $\frac{1}{2}$
   (d) $\frac{1}{2}$ or an odd integral multiple of $\frac{1}{2}$

37. The chemical potential of a photon gas is
   (a) zero
   (b) positive
   (c) negative
   (d) imaginary

38. The chemical potential of a fermionic system with a fixed number of particles at absolute zero of temperature is
   (a) zero
   (b) negative
   (c) positive
   (d) imaginary

39. The classical statistical mechanics will give accurate results for a gas of free molecules, when the mean separation $(\bar{r})$ between the

molecules and their thermal de Broglie wavelength (λ) have the following relationship

(a) $\lambda \gg \vec{r}$
(b) $\lambda = \vec{r}$
(c) $\lambda \ll \vec{r}$
(d) none of the above

40. A micro-canonical ensemble represents
    (a) a system in contact with a heat reservoir
    (b) an isolated system in equilibrium
    (c) a system that can exchange particles with its surroundings
    (d) a system under constant external pressure

41. Suppose temperature of the sun goes down by a factor of two, then the total power emitted by the sun will go down by a factor of
    (a) 2
    (b) 4
    (c) 8
    (d) 10

42. An ideal gas in a cylinder is compressed adiabatically to one-third of its initial volume. During this process, 20 J work is done on the gas by compressing agent. Which of the following statements is true for this case?
    (a) Change in internal energy in this process is zero.
    (b) The internal energy increased by 20 J.
    (c) The internal energy decreased by 20 J.
    (d) Temperature of the gas decreases.

43. Mean total energy of a classical three-dimensional harmonic oscillator in equilibrium with a heat reservoir at temperature T is
    (a) $k_B T$
    (b) $\frac{3}{2} k_B T$
    (c) $2 k_B T$
    (d) $3 k_B T$

44. Which of the following is not an exact differential?
    (a) dQ (Q = heat absorbed)
    (b) du (u = internal energy)
    (c) dS (S = entropy)
    (d) dF (F = free energy)

45. The root mean square speed of a particle of mass m in the kinetic theory is given by
    (a) $\sqrt{\dfrac{k_B T}{m}}$
    (b) $\sqrt{\dfrac{2 k_B T}{m}}$
    (c) $\sqrt{\dfrac{3 k_B T}{m}}$
    (d) $\sqrt{\dfrac{8 k_B T}{m}}$

46. The temperature of a wire vibrating between two fixed supports will
    (a) remain same
    (b) becomes zero
    (c) drop
    (d) increase if tension is increased

47. An ideal gas is one in which there is
    (a) strong interaction
    (b) no interaction
    (c) weak interaction
    (d) medium interaction between the particles

48. At equilibrium the entropy of an isolated system having energy between u and $u+\delta u$ is
    (a) $\ln \Omega(u,v,w)$
    (b) $k \ln \Omega(u,v,w)$
    (c) $-k \ln \Omega(u,v,w)$
    (d) $-\ln \Omega(u,v,w)$

49. If the phase space is divided into cells, each of size h^f, then set of microscopic states contained in a volume element $\Delta \Gamma$ corresponds to a set of
    (a) $\dfrac{\Delta \Gamma}{h^f}$
    (b) $\Delta \Gamma h^f$
    (c) $\Delta \Gamma + h^f$
    (d) $\Delta \Gamma - h^f$

50. The differential form of the first law of thermodynamics for a non-diffusively interacting system is
    (a) $du = \delta Q + \delta W$
    (b) $du = \delta Q - \delta W$
    (c) $du = \delta Q - \delta W + \mu dN$
    (d) $du = \delta Q + \delta W + \mu dN$

51. System consisting of identical particles of half-odd integral spin $\dfrac{\hbar}{2}, \dfrac{3\hbar}{2}, \dfrac{5\hbar}{2},\ldots$ are described by
    (a) Bose–Einstein statistics
    (b) symmetric wave function $\psi(S)$
    (c) antisymmetric wave function $\psi(A)$
    (d) Maxwell's–Boltzmann statistics

52. Helmoltz free energy of system A in the surroundings of constant temperature T^v is
    (a) $-T \ln Z$
    (b) $u + T^v S$
    (c) $u - T^v S$
    (d) $kT \ln Z$

53. System consisting of identical particles of integral spin $0, 1\hbar, 2\hbar,...$ are described by
    (a) antisymmetric wave functions
    (b) symmetric wave functions
    (c) Fermi–Dirac statistics
    (d) Maxwell's–Boltzmann statistics

54. One mole of a gas, assumed to be perfect at 0°C is heated at constant pressure till its volume is twice its initial value. The amount of heat absorbed is $(C_v = 20.9 \text{ J mol}^{-1}\text{K}^{-1})$, $(R = 8.3 \text{ J mol}^{-1}\text{K}^{-1})$
    (a) $0.797 \times 10^2 \text{J}$
    (b) $79.7 \times 10^2 \text{J}$
    (c) $7.97 \times 10^2 \text{J}$
    (d) $797 \times 10^2 \text{J}$

55. An operator whose expectation value for all admissible wave functions is real is known as
    (a) Lagrangian operator
    (b) Hermitian operator
    (c) Skew Hermitian operator
    (d) Laplacian operator

56. For a reversible isothermal and isobaric process
    (a) $\Delta S = 0$
    (b) $\Delta S = \text{constant}$
    (c) $\Delta G = 0$
    (d) $\Delta G = \text{constant}$

57. The entropy σ of a system in statistical equilibrium is defined by
    (a) $\sigma(V_1 E) = \Delta \Gamma(E)$
    (b) $\sigma(V_1 E) = \dfrac{1}{\Delta \Gamma(E)}$
    (c) $\sigma(V_1 E) = \ln \Delta \Gamma(E)$
    (d) $\sigma(V_1 E) = \dfrac{1}{\ln \Delta \Gamma(E)}$

58. Consider a system that may be either unoccupied or occupied by one particle with energy 0 or E. The absolute activitiy of the system is given by
    (a) $Z(\mu, T) = e^{\mu B}$
    (b) $Z(\mu, T) = e^{\frac{1}{kT}}$
    (c) $Z(\mu, T) = e^{\mu}$
    (d) $Z(\mu, T) = e^{kT}$

59. The thermal de Broglie wavelength associated with the molecule of a gas at temperature T is

(a) $h2\pi mkT$

(b) $\dfrac{h}{2\pi mkT}$

(c) $h(2\pi mkT)^{\frac{1}{2}}$

(d) $\dfrac{h}{(2\pi mkT)^{\frac{1}{2}}}$

60. The vapour pressure of solid ammonia is

(a) $\ln p = 23.03 - \dfrac{3754}{T}$

(b) $\ln p = 23.03 + \dfrac{3754}{T}$

(c) $\ln p = 19.49 - \dfrac{3063}{T}$

(d) $\ln p = 19.49 + \dfrac{3063}{T}$

61. The mixing of two different gases is an irreversible process. It is therefore attended by

(a) constant entropy
(b) decrease of entropy
(c) zero entropy
(d) an increase in entropy

62. The total rate of change of density $\dfrac{dP}{dt}$ in the vicinity of any selected phase point of a system as it moves through the P space

(a) increases
(b) decreases
(c) is zero
(d) is constant

63. According to Nernst postulate, the entropy of any system vanishes in the state for which

(a) $T = 0$ K
(b) $I = 4$ K
(c) $T = 298.16$ K
(d) $T = 77$ K

64. The point in phase space is actually a cell whose minimum volume is of the order of

(a) h
(b) h^3
(c) h^2
(d) h^5

65. The value of the thermal de Broglie wavelength λ in Å for an electron at room temperature is

(a) $0.745 T^{-\frac{1}{2}}$
(b) $0.745 T^{\frac{1}{2}}$
(c) $745 \, T^{-\frac{1}{2}}$
(d) $745 \, T^{\frac{1}{2}}$

66. In bosons, the spin of the nucleus is
    (a) zero only
    (b) an integral number
    (c) either zero or an integral number
    (d) an odd integral multiple of $\dfrac{1}{2}$

67. $G = U - T's + p'V$ is Gibbs free energy of system A in an environment of
    (a) constant entropy S and constant pressure p'
    (b) constant temperature T' and constant entropy S
    (c) constant temperature T' and constant pressure p'
    (d) constant volume V and constant pressure p'

68. Classical statistical mechanics will give accurate results only in those cases where the energy difference between the quantum states is less than
    (a) $h\upsilon$ (b) $\dfrac{h\upsilon}{kT}$ (c) e^{kT} (d) kT

69. For a reversible isothermal and isobaric process
    (a) $\Delta N = 0$ (b) $\Delta G = 0$ (c) $\Delta G \neq 0$ (d) $\Delta u = 0$

70. For a monoatomic noble gas $C_v = \left(\dfrac{dE}{dT}\right)_V$ is
    (a) 0.1247 J/mole-K
    (b) 1.247 J/mole-K
    (c) 12.47 J/mole-K
    (c) 124.7 J/mole-K

71. The expression for calculating the relative population between any two energy states is
    (a) $\dfrac{n_i}{n_j} = -\dfrac{\Delta E_{ij}}{RT}$
    (b) $\dfrac{n_i}{n_j} = \exp[RT]$
    (c) $\dfrac{n_i}{n_j} = \exp\left[\dfrac{\Delta E_{ij}}{RT}\right]$
    (d) $\dfrac{n_i}{n_j} = \exp\left[-\dfrac{\Delta E_{ij}}{RT}\right]$

72. The work function ϕ of the metal is
    (a) $\mu_B T\, \mu(T)$
    (b) $\mu_B - \mu(T)$
    (c) $\mu_B \times \mu(T)$
    (d) $\dfrac{\mu_B}{\mu(T)}$

73. Richardson–Dushmann equation for thermionic emission is
    (a) $J = A'T^2 e^{-\frac{\phi}{kT}}$
    (b) $J = A'T^4 e^{-\frac{\phi}{kT}}$
    (c) $J = A'T^2$
    (d) $J = T^4 e^{-\frac{\phi}{kT}}$

74. The most spectacular effect of superfluidity is
    (a) high viscosity
    (b) finite thermal conductivity
    (c) second sound
    (d) capillarity

75. The lowest value of partition function at absolute zero when all particles occupy the lowest energy state is
    (a) zero   (b) 1   (c) -1 (d) 10

76. At atmospheric pressure, helium condenses into a normal liquid at
    (a) 4.2 K   (b) 0.42 K   (c) 3 K   (d) 3.2 K

77. The Fermi energy of electron gas in copper at 300 K is
    (a) 0.7 eV   (b) 0.07 eV   (c) 7.0 eV   (d) 70 eV

78. The number of microstates accessible to a system should increase with
    (a) increase in its energy
    (b) decrease in its energy
    (c) decrease in entropy
    (d) decrease in ensemble

79. In Bose–Einstein statistics, the chemical potential is always
    (a) zero   (b) positive   (c) infinity   (d) negative

80. When two systems at different temperatures are put into thermal conductivity, the total entropy
    (a) remains same
    (b) becomes zero
    (c) decreases
    (d) increases

81. The pressure of black body radiation at 6000 K is
    (a) 0.326 N/m^2
    (b) 3.26 N/m^2
    (c) 32.6 N/m^2
    (d) 326 N/m^2

82. When a system consisting of mobile magnetic particles is placed in a non-uniform magnetic field, the particle density in a region of higher magnetic field
    (a) remains constant
    (b) increases
    (c) decreases
    (d) becomes zero

83. The second sound is a temperature or
    (a) pressure wave
    (b) transverse wave
    (c) entropy wave
    (d) stationary wave

84. According to the equipartition theorem, when a system is in thermal equilibrium with a heat bath at temperature T, the mean contribution of each quadratic term to the total energy is
    (a) $\dfrac{kT}{2}$
    (b) $\dfrac{3kT}{2}$
    (c) kT
    (d) $\dfrac{1}{2}kT$

85. When a magnetic field B is applied, the energy of electrons in the D state is greater than the energy of electrons in the U state by an amount
    (a) μ_B^2
    (b) $2\mu_B^2 B$
    (c) $2\mu_B B$
    (d) $\mu_B B$

86. A box of volume 2V with thermally insulating walls is partitioned into two equal compartments one of which contains an ideal gas at temperature T and the other is empty. The partition is suddenly removed. The temperature of the gas finally will be
    (a) T
    (b) $\dfrac{T}{2}$
    (c) $\dfrac{T}{4}$
    (d) $2T$

87. For a classical gas of particles each of mass m in thermal equilibrium at temperature T, the mean values of V_x and of $(V_y^2 + V_z^2)$ are
    (a) $\left(\sqrt{\dfrac{1}{3}}\right)C$ and $\left(\dfrac{2}{3}\right)C^2$
    (b) $\left(\dfrac{1}{3}\right)C^2$ and $\left(\dfrac{2}{3}\right)C$
    (c) $\left(\sqrt{\dfrac{1}{3}}\right)C^2$ and $\left(\sqrt{\dfrac{2}{3}}\right)C^2$
    (d) $\left(\dfrac{1}{3}\right)C$ and $\left(\sqrt{\dfrac{2}{3}}\right)C$

    where, quantities V_x, V_y and V_z are the components of the velocity of a gas particle.

88. In deriving Planck's radiation formula, the photons are heated as
    (a) Fermions   (b) Bosons
    (c) Baryons   (d) Photons

89. Which law of thermodynamics states that the entropy of a system vanishes at absolute zero?
    (a) zeroth law   (b) first law
    (c) second law   (d) third law

90. At short wavelength, Planck's radiation formula reduces to
    (a) Wien's law   (b) Stefan's law
    (c) Rayleigh–Jean's law (c) Kirchoff's law

91. If x represents the square of the velocity of a molecule and if the number of molecules in a gas having velocity v, then according to Maxwell–Boltzmann statistics, the plot of x and y is
    (a) a curve rising with x
    (b) a curve falling with x
    (c) a straight line parallel to x-axis
    (d) a curve with a maximum

92. A material with a constant heat capacity 10 J/K is heated from temperature 27°C to 127°C, the change in its entropy is

    (a) 19.7 J/K   (b) $10\ln\left(\frac{4}{3}\right)$ J/K

    (c) $\ln(0.75)$ J/K   (d) 2.3 J/K

93. For a diatomic ideal gas, the ratio of specific heat $\frac{C_p}{C_v}$ is equal to

    (a) $\frac{5}{3}$ (b) $\frac{4}{3}$ (c) $\frac{1}{2}$ (d) $\frac{7}{5}$

94. The ratio of the average velocities of oxygen molecules and nitrogen molecules at NTP in air is

    (a) $\frac{7}{8}$ (b) $\left(\frac{7}{8}\right)^{\frac{1}{2}}$ (c) 1.005   (d) 0.82

95. The Helmholtz free energy of an ideal gas is

$$F = NRT\left[\frac{f_0 - \ln T^{\frac{3}{2}}V}{N}\right],$$ where, N = mole number, V = volume, t = temperature and f_0 is a constant, the entropy S is given by

(a) $-NR\left[\frac{f_0 - \ln T^{\frac{3}{2}}V}{N}\right] + \frac{3}{2}NR$ (b) $\frac{3}{2}NR$

(c) $\frac{3f_0}{2}NR$ (d) $NR\left[f_0 - \frac{3}{2}\ln\frac{V}{N}\right]$

96. Regarding Bose–Einstein condensation of an ideal Bose gas, which of the following statements is correct?

(a) The Bose–Einstein condensation can be observed only for zero particles like photons.

(b) The Bose–Einstein condensation occurs at densities lower than a threshold density p_c.

(c) The Bose–Einstein condensation does not occur in two dimensions.

(d) To observe Bose–Einstein condensation, the electron density of the material should be very high.

97. If C_v is the specific heat of a solid at temperature T, at very low temperatures, the slope of the $\log C_v - \log T$ plot is

(a) 0.5   (b) 1   (c) 2   (d) 3

98. Heat sinks are used in power amplifiers to

(a) increase output power

(b) reduce heat losses in the transistor

(c) increase voltage gain

(d) increase collector dissipation rating of the transistor

99. According to Maxwell's velocity distribution law, the number of particles with speed V is proportional to

(a) Ve^{-CV} (b) $V^2 e^{-CV^2}$ (c) Ve^{-CV^2} (d) $V^2 e^{-CV}$

100. For an ideal Fermi-gas, the function (E_F = Fermi energy, V = volume per particle) represents
   (a) C_v
   (b) Helmholtz free energy
   (c) zero point energy
   (d) zero point pressure

101. Which of the following represents a first-order phase transition?
   (a) onset of ferro electricity
   (b) superconductivity
   (c) order–disorder transition in alloys
   (d) sublimation

102. If ΔS is the change in the entropy of the universe, which of the following is correct?
   (a) $\Delta S > 0$ (b) $\Delta S = 0$ (c) $\Delta S < 0$ (d) $\Delta S \neq 0$

103. Which law of thermodynamics states that the specific heat of a substance approaches zero as the temperature approaches absolute zero?
   (a) zeroth law
   (b) first law
   (c) second law
   (d) third law

104. Debye's theory predicts that at very low temperatures, the specific heat of a solid is proportional to
   (a) T
   (b) T^2
   (c) T^3
   (d) T^{-3}

105. A system is said to be in thermodynamic equilibrium if it is in
   (a) mechanical equilibrium
   (b) thermal equilibrium
   (c) chemical equilibrium
   (d) mechanical, thermal and chemical equilibrium

106. At low speeds, the plot of number of particles in a classical gas and their speed is
   (a) a straight line
   (b) a parabola
   (c) a hyperbola
   (d) an exponential curve

107. In a system of Fermions, with fermi temperature T_F, the distribution is strongly degenerate when
   (a) $T \ll T_F$ (b) $T > T_F$ (c) $T \gg T_F$ (d) $T < T_F$

108. The thermal expansion of a one-dimensional lattice will be zero if the P.E. is of the form
    (a) Cx^2
    (b) $Cx^2 - Dx^2$
    (c) $Cx^2 - Dx^2 - Fx^5$
    (d) $Cx^2 - Fx^5$

109. According to the free electron theory, the resistivity of a metal at high temperature is
    (a) directly proportional to T
    (b) inversely proportional to T
    (c) independent of T
    (d) directly proportional to T^2

110. Which of the following statements is not due to the second law of thermodynamics?
    (a) There are no perfect engines.
    (b) There are no perfect refrigerators.
    (c) In an isolated system, the entropy must increase.
    (d) In an isolated system, changes in which the entropy decreases will not happen.

111. According to Liouville's theorem
    (a) the density of systems in the neighbourhood of some given system in coordinate space remains invariant in time.
    (b) the density of systems in the neighbourhood of some given system in momentum space does not change with time.
    (c) the density of systems in the neighbourhood of some given system in phase space remains constant in time.
    (d) the characteristics of the ensemble is independent of the density function.

112. Pick out the correct statement concerning ensembles.
    (a) In micro-canonical ensemble, the density function is not a Dirac delta function involving the energy.
    (b) In a canonical-ensemble, the subsystem could exchange particles, but not energy with the reservoir.
    (c) In ground-canonical ensemble, the subsystem can exchange energy but not particles with the reservoir.
    (d) In ground-canonical ensemble, the subsystem can exchange energy, as well as particles with the reservoir.

113. For an ideal Bose gas
    (a) the chemical potential is always positive
    (b) the chemical potential is never negative
    (c) the chemical potential is never positive
    (d) the chemical potential can be positive, zero or negative

114. In an ideal Fermi gas
    (a) the pressure is non-zero even at absolute zero
    (b) the chemical potential is positive at high temperatures and negative at low temperatures
    (c) the distribution function has the value 0.33 at the Fermi level
    (d) the occupancy of the states with energy greater than the Fermic energy is non-zero at absolute zero

115. Choose the correct statement concerning the paramagnetic susceptibility of Fermi gas conduction electrons at very low temperatures.
    (a) it varies at $\frac{1}{T}$
    (b) it varies as T
    (c) it is independent of T
    (d) classical theory can beautifully account for it

116. A Carnot engine absorbs 100 calories of heat from a source at 400 K and gives up 80 calories for a sink. The temperature of the sink is
    (a) 20 K
    (b) 300 K
    (c) 320 K
    (d) 500 K

117. The ensemble used to obtain the equation of state for an ideal Bose gas is
    (a) canonical ensemble
    (b) micro-canonical ensemble
    (c) macro-canonical ensemble
    (d) grand-canonical ensemble

118. For a super fluid _____ is zero
    (a) surface tension
    (b) viscosity
    (c) density
    (d) specific heat

119. The internal energy of an electron gas is proportional to

(a) T (b) $\dfrac{1}{T}$ (c) T^2 (d) $\dfrac{1}{T^2}$

120. Planck's law becomes classical equation at
   (a) low temperatures and long wavelengths
   (b) low temperatures and short wavelengths
   (c) high temperatures and short wavelengths
   (d) high temperatures and long wavelengths

121. If $\overline{V}$, V_p and V_{rms} denote the average, most probable and root mean square values respectively of the molecular speeds for a gas at room temperature, obeying Maxwellian velocity distribution, then

(a) $V_{rms} < \overline{V} < V_p$ (b) $V_{rms} < V_p < \overline{V}$

(c) $\overline{V} < V_{rms} < V_p$ (d) $V_p < \overline{V} < V_{rms}$

122. For an isolated thermodynamical system, P, V, T, U, S and F represent the pressure, volume, temperature, internal energy, entropy and free energy respectively. Then the following relation is true

(a) $\left(\dfrac{\partial F}{\partial T}\right)_V = -S$ (b) $\left(\dfrac{\partial F}{\partial T}\right)_P = -S$

(c) $\left(\dfrac{\partial V}{\partial S}\right)_P = T$ (d) $\left(\dfrac{\partial U}{\partial V}\right)_T = -P$

123. Thermal expansion in solids with increasing temperature is a consequence of
   (a) anharmonicity of the lattice vibration
   (b) pressure of the electron gas
   (c) dislocations in the lattice
   (d) none of the above

124. The ratio $\dfrac{\mu_{Cu}}{\mu_{Si}}$ —the electron mobility in copper to that in—silicon is approximately

(a) $\dfrac{1}{30}$ (b) $\dfrac{1}{3}$ (c) 3 (d) 30

125. The sole effect of a hypothetical thermodynamic process E_1 is to convert an amount of heat entirely into work. A second hypothetical process E_2 converts an amount W of work entirely into heat. It can be said that the second law of thermodynamics is violated by

(a) E_1 but not E_2
(b) E_2 but not E_1
(c) both E_1 and E_2
(d) neither process

126. A carbon rod is heated. Its resistance

(a) increases
(b) decreases
(c) remains the same
(d) varies sinusoidally

127. In an ideal gas at a given temperature, the pressure exerted by a gas is

(a) inversely proportional to the density of the gas
(b) independent of the density of the gas
(c) directly proportional to the square of the density of the gas
(d) directly proportional to the density of the gas

128. The absolute zero of temperature is that temperature at which

(a) water freezes
(b) nitrogen is a liquid
(c) pressure of all gases is zero
(d) ice melts

129. The temperature of an ideal gas is decreased from 300 K to 150 K. If the thermal velocity at the original temperature is 40,000 m/s, what is the velocity at the lower temperature?

(a) 160,000 m/s
(b) 56,000 m/s
(c) 28,300 m/s
(d) 20,000 m/s

130. During the refrigeration cycle, latent heat of condensation takes place in the

(a) evaporator
(b) condensor
(c) metering device
(d) compressor

131. The Widemann and Franz ratio is

(a) inversely proportional to absolute temperature
(b) directly proportional to square root of absolute temperature
(c) directly proportional to absolute temperature
(d) inversely proportional to square of the temperature

132. If $\bar{\omega}$ is the thermodynamic probability of the state of the system, then the entropy of the system is

   (a) $k \cdot \omega$ (b) $k \exp(\omega)$ (c) $k \log(\omega)$ (d) $\log \dfrac{\omega}{k}$

133. The particles whose spin is an integral multiple of h are called as

   (a) fermions   (b) leptons   (c) bosons   (d) mesons

134. The phenomenon of expulsion of magnetic flux in a superconductor is called as

   (a) Cooper effect
   (b) Meissner effect
   (c) BCS effect
   (d) Zeeman effect

135. The ratio of average kinetic energies of molecules of gas when the temperature is raised from 27°C to 177°C is

   (a) 3 : 2   (b) 2 : 3   (c) 1 : 2   (d) 2 : 1

136. A refrigerator works between 3°C and 40°C. To keep the temperature of the refrigerator space constant, 600 calories of heat are to be removed every second. The power required is (J = 4.2 Joules/cal)

   (a) 23.65 Watt
   (b) 236.5 Watt
   (c) 2365 Watt
   (d) 2.365 Watt

137. One should 'Defrost' the refrigerator because

   (a) it adds unnecessary extra weight
   (b) ice is a good conductor of heat
   (c) ice is a poor conductor of heat
   (d) ice has less heat capacity than water

138. A steam engine takes steam from the boiler at 200°C and exhausts directly into the air at 100°C. What is its maximum possible efficiency?

   (a) 2.11%   (b) 21.1%   (c) 22%   (d) 40%

139. The root mean square velocity of molecules of a gas is given by the expression

   (a) $1.73\sqrt{\dfrac{kT}{m}}$ (b) $1.73\sqrt{\dfrac{m}{kT}}$

   (c) $1.5\sqrt{\dfrac{kT}{m}}$ (d) $1.5\sqrt{\dfrac{m}{kT}}$

140. In Maxwell's law of distribution of velocities
    (a) the number of molecules with most probable velocity is constant
    (b) the number of molecules with most probable velocity is small
    (c) the number of molecules with most probable velocity is large
    (d) the number of molecules with most probable velocity is infinite

141. Two samples of gas initially at same temperature and pressure are compressed to half their original volume are isothermally and the other adiabatically, the final pressure will be
    (a) greater for adiabatic compression
    (b) lesser for adiabatic compression
    (c) greater for isothermal compression
    (d) same for both the cases

142. Consider a particle of mass m at temperature T which follows classical Boltzmann statistics. The average speed is given by
    (a) $\sqrt{\dfrac{8kT}{\pi m}}$ (b) $\sqrt{\dfrac{6kT}{\pi m}}$ (c) $\sqrt{\dfrac{4kT}{\pi m}}$ (d) $\sqrt{\dfrac{2kT}{\pi m}}$

143. The specific heat of metallic copper (over the temperature range 15 to 100°C) is approximately 0.093 cal gm^{-1} deg^{-1} (one calorie $= 4.184 \times 10^7$ ergs). Neglecting thermal expansion, the heat capacity of 10 g of specimen is given by
    (a) 9.3×10^6 ergs deg^{-1} (b) 3.80×10^7 ergs deg^{-1}
    (c) 3.89×10^6 ergs deg^{-1} (d) 4.184×10^7 ergs deg^{-1}

144. For a pure substance with one component, the number of phases that can exist in contact cannot exceed
    (a) 2   (b) 3   (c) 4   (d) 5

145. The maximum of the black body energy density $E(\omega)$ as a function of ω is characterized by
    (a) $\lambda_{max} T = \dfrac{hc}{2.82k}$ (b) $\lambda_{max} = \dfrac{hc}{3Tk}$
    (c) $\lambda_{max} = \dfrac{hc}{kT^{\frac{3}{2}}}$ (d) $\lambda_{max} = \dfrac{hc}{3.1k}$

146. We can define the temperature T as follows
    (a) the reciprocal of the temperature is equal to the derivative of the entropy with respect to energy
    (b) the temperature is the derivative of energy with respect to entropy
    (c) the temperature is the derivative of entropy with respect to energy
    (d) the reciprocal of the temperature is the derivative of energy with respect to entropy

147. For a gas with van der Waals' equation of state $\left[P + \dfrac{a}{V^2}\right](V-b) = RT$, the coefficient of cubical expansion β is given by

    (a) $\beta = \dfrac{RV^2(V-b)}{\left[RTV^3 - 2a(V-b)^2\right]}$

    (b) $\beta = \dfrac{1}{kT}$

    (c) $\beta = \dfrac{RV^2(V-a)}{\left[RTV^3 - 2b(V-a)^2\right]}$

    (d) $\beta = RV^2 V(1-b)(V-a)$

148. For a Fermi gas of electrons with low non-zero temperatures, the variation of thermal energy with respect to temperature is [$E_F(e)$ = electron Fermi energy]

    (a) $\dfrac{T^2}{E_F(e)}$
    (b) $\dfrac{T}{E_F(e)}$
    (c) $\dfrac{\sqrt{T}}{E_F(e)}$
    (d) $E_F(e)T^{\frac{3}{2}}$

149. One-dimensional Ising model
    (a) shows ferromagnetism at $T = 0$
    (b) shows ferromagnetism for $T < T_c$
    (c) shows ferromagnetism for $T > T_c$
    (d) never exhibits ferromagnetism

150. The specific heat C_V is proportional to T^3 and is not bounded as $T \to \infty$ in Planck's theory of radiation due to the fact that
    (a) photons have zero rest mass
    (b) photons have two transverse polarizations
    (c) the number of photons in the cavity is not bounded
    (d) the number of photons in the cavity is bounded

151. Bose–Einstein condensation temperature is defined as the temperature at which
    (a) the number of atoms in the excited state is equal to the total number of atoms
    (b) the number of atoms in the ground state equals the number of atoms in the excited state
    (c) the number of atoms in the excited state is 0
    (d) the ground state becomes unstable

ANSWERS

1. (c)	2. (a)	3. (a)	4. (d)	5. (c)
6. (b)	7. (c)	8. (c)	9. (c)	10. (b)
11. (b)	12. (d)	13. (a)	14. (a)	15. (c)
16. (b)	17. (c)	18. (b)	19. (a)	20. (b)
21. (a)	22. (b)	23. (a)	24. (d)	25. (d)
26. (d)	27. (c)	28. (d)	29. (c)	30. (b)
31. (b)	32. (c)	33. (d)	34. (b)	35. (b)
36. (d)	37. (a)	38. (c)	39. (c)	40. (b)
41. (c)	42. (b)	43. (b)	44. (a)	45. (c)
46. (c)	47. (b)	48. (b)	49. (a)	50. (a)
51. (c)	52. (c)	53. (b)	54. (b)	55. (b)
56. (c)	57. (c)	58. (a)	59. (d)	60. (a)
61. (d)	62. (c)	63. (a)	64. (b)	65. (c)
66. (c)	67. (c)	68. (d)	69. (b)	70. (c)
71. (d)	72. (b)	73. (a)	74. (c)	75. (b)
76. (a)	77. (c)	78. (a)	79. (d)	80. (c)
81. (a)	82. (b)	83. (c)	84. (a)	85. (c)

86. (a)	87. (a)	88. (d)	89. (d)	90. (a)
91. (d)	92. (d)	93. (d)	94. (b)	95. (a)
96. (b)	97. (d)	98. (b)	99. (b)	100. (c)
101. (d)	102. (a)	103. (c)	104. (c)	105. (d)
106. (d)	107. (c)	108. (a)	109. (b)	110. (c)
111. (c)	112. (d)	113. (c)	114. (d)	115. (c)
116. (d)	117. (a)	118. (b)	119. (c)	120. (d)
121. (d)	122. (a)	123. (a)	124. (d)	125. (a)
126. (b)	127. (d)	128. (c)	129. (c)	130. (b)
131. (c)	132. (c)	133. (c)	134. (b)	135. (d)
136. (b)	137. (c)	138. (b)	139. (a)	140. (b)
141. (a)	142. (a)	143. (b)	144. (a)	145. (a)
146. (b)	147. (d)	148. (a)	149. (d)	150. (c)
151. (a)				

3

MATHEMATICAL PHYSICS

IMPORTANT VECTOR IDENTITIES

1. $\nabla(\phi+\psi) = \nabla\phi + \nabla\psi$
2. $\nabla(\phi\psi) = \phi\nabla\psi + \psi\nabla\phi$
3. div $(A + B)$ = div A + div B
4. curl $(A + B)$ = curl A + curl B
5. $\text{grad}(A \cdot B) = A \times (\nabla \times B) + (A \cdot \nabla)B + B \times (\nabla \times A) + (B \cdot \nabla)A$
6. div$(\phi A) = \phi$ div $A + A \cdot \text{grad}\,\phi$
7. curl$(\phi A) = \phi$ curl $A + \text{grad}\,\phi \times A$
8. div curl $A = 0$
9. div$(A \times B) = B \cdot \text{curl}\,A - A \cdot \text{curl}\,B$
10. curl$(A \times B) = (B \cdot \nabla)A - (A \cdot \nabla)B + A$ div$B - B$ div$A$
11. curl curl$A = \text{grad div } A - \nabla^2 A$

    where, ϕ and ψ are scalar point functions and A, B are vector point functions in a certain region.

12. **Gauss divergence theorem** The flux of a vector A over any closed surface S is equal to the volume integral of the divergence of the vector field over the volume enclosed by the surface S. That is,

$$\iint_S A \cdot ds = \iiint_V \text{div}\,A\, dV$$

13. **Stoke's theorem** The flux of the curl of a vector function A over surface S of any shape is equal to the line integral of the vector A over the boundary C of the surface and is given by

$$\iint_S \operatorname{curl} A \cdot ds = \int_C A \cdot dr$$

14. **Green's theorem in a plane** If S is a closed surface of XY plane bounded by the closed curve C and if M and N are continuous functions of x and y having continuous derivatives in S, then

$$\oint_C M\,dx + N\,dy = \iint_S \left(\frac{\partial N}{\partial x} - \frac{\partial M}{\partial y} \right) dx\,dy$$

MATRICES

15. **Transpose of a matrix** A matrix of order $n \times m$ obtained by interchanging the rows and columns of $(m \times n)$ matrix A is called the transpose of A and is denoted as A' or $\tilde{A}$ or A^T.

    Properties of transpose matrix

    (a) $(A^T)^T = A$
    (b) $(\lambda A)^T = \lambda A^T$, λ being any scalar
    (c) $(A+B)^T = A^T + B^T$
    (d) $(AB)^T = B^T A^T$

16. **Conjugate of a matrix** If A is any matrix having complex numbers, then the matrix obtained from A by replacing its each element by its conjugate complex number is called the conjugate of matrix A and is denoted by $\bar{A}$ or A^*. Thus if $A = [a_{ij}]$ then $A^* = [a_{ij}]^*$.

    Properties

    (a) $(A^*)^* = A$
    (b) $(A+B)^* = A^* + B^*$
    (c) $(\lambda A)^* = \lambda^* A^*$
    (d) $(AB)^* = A^* B^*$

17. **Conjugate transpose of a matrix** Conjugate of the transpose of a matrix A is called conjugate transpose of A and is denoted by A^+ or $(A^T)^*$.

    Properties

    (a) $(A^+)^+ = A$
    (b) $(A+B)^+ = A^+ + B^+$
    (c) $(\lambda A)^+ = \lambda^* A^+$
    (d) $(AB)^+ = B^+ A^+$

18. **Symmetric and antisymmetric matrices** A square matrix $A = [a_{ij}]$ is said to be symmetric provided $a_{ij} = a_{ji}$ for all values of i and j.

    i.e., $$A^T = A$$

    A square matrix is said to be antisymmetric or Skew symmetric provided $a_{ij} = -a_{ji}$

    i.e., $$A^+ = -A$$

19. **Hermitian and Skew-Hermitian matrices** A square matrix $A = [a_{ij}]$ is said to be Hermitian matrix if $a_{ij} = a_{ji}^*$

    i.e., $$A^+ = A$$

    A square matrix $A = [a_{ij}]$ is said to be Skew-Hermitian matrix if $a_{ij} = -a_{ji}$

    i.e., $$A^+ = -A$$

20. **Orthogonal matrices**

    $$A^T A = I$$

    $$A A^T = I$$

21. **Unitary matrices**

    $$A A^+ = I$$

BETA, GAMMA AND ERROR FUNCTIONS

22. Beta function is defined by

    $$\beta(m, n) = \int_0^1 x^{m-1}(1-x)^{n-1} dx \quad \begin{cases} m > 0 \\ n > 0 \end{cases}$$

23. Gamma function is defined by

    $$\Gamma_n = \int_0^\infty e^{-x} x^{n-1} dx, \quad n > 0$$

24. Error function or the probability integral is defined by

$$erf(x) = \frac{2}{\sqrt{\pi}} \int_0^x e^{-y^2} dy$$

25. Symmetry property of Beta function is

$$\beta(m, n) = \beta(n, m)$$

$$\beta(m, n) = \int_0^1 x^{m-1}(1-x)^{n-1} dx \quad \begin{cases} m > 0 \\ n > 0 \end{cases}$$

Substituting $x = 1 - y$, $dx = -dy$ in above equation, we get

$$\beta(m, n) = \int_0^1 (1-y)^{m-1} y^{n-1} dy$$

$$= \int_0^1 x^{n-1}(1-x)^{m-1} dx$$

$$= \beta(n, m)$$

26. Relation between beta and gamma function is

$$\beta(m, n) = \frac{\Gamma m \, \Gamma n}{\Gamma(m+n)}$$

COMPLEX VARIABLES

27. The Maclaurin expansions of $\cos\theta$, $\sin\theta$ and e^t is given as

$$\cos\theta = 1 - \frac{\theta^2}{2!} + \frac{\theta^4}{4!} - \frac{\theta^6}{6!} + ...$$

$$\sin\theta = \theta - \frac{\theta^3}{3!} + \frac{\theta^5}{5!} - \frac{\theta^7}{7!} + ...$$

$$e^t = 1 + \frac{t}{1!} + \frac{t^2}{2!} + \frac{t^3}{3!} + ...$$

28. **Basic properties of the complex integrals** If $f_1(z)$ and $f_2(z)$ are continuous complex functions over a continuous rectifiable curve C so that $f_1(z)$ and $f_2(z)$, $f_1(z) + f_2(z)$ are all integrable over C, then,

   i. $\int_C [f_1(z) + f_2(z)] dz = \int_C f_1(z) dz + \int_C f_2(z) dz$

   ii. If C is written as C_1 and C_2, then
   $$\int_C f(z) dz = \int_{C_1} f(z) dz + \int_{C_2} f(z) dz$$

   iii. If we reverse the sense of integration, the sign of the integral value changes is
   $$\int_C f(z) dz = -\int_{-C} f(z) dz$$
   Here, $-C$ represents the sense opposite to that of C.

   iv. If K is any constant, then,
   $$\int_C K f(z) dz = K \int_C f(z) dz$$

29. **Cauchy's integral theorem** If $f(z)$ is analytic throughout a simply connected bounded domain D, then for every closed contour C in the domain D
$$\int_C f(z) dz = 0$$

30. **Cauchy's integral formula** If $f(z)$ is analytic and single-valued within and on a closed contour C and if z_0 is any point within C then,
$$f(z_0) = \frac{1}{2\pi i} \int_C \frac{f(z)}{z - z_0} dz$$

31. **Liouville's theorem** If $f(z)$ is analytic and bounded in absolute value for all (finite) of z in the complex plane, then $f(z)$ is a constant.
$$f^n(z) = \frac{n!}{2\pi i} \int_C \frac{f(z) dz}{(z - z_0)^{n+1}} \qquad (n = 1, 2, 3,...)$$

32. **Taylor's series** If $f(z)$ is analytic at all points inside a circular domain D with its centre at $z = z_0$ and radius r_0 then for every z inside D

$$f(z) = f(z_0) + \frac{f'(z_0)}{1!}(z, z_0) + \frac{f''(z_0)}{2!}(z - z_0)^2 + \ldots$$

$$+ \frac{f^{(n)}(z_0)}{n!}(z - z_0)^n + \ldots$$

33. A linear differential equation of order n is of the form

$$f(x) = a_0 \frac{d^n y}{dx^n} + a_1(x) \frac{d^{n-1}}{dx^{n-1}} + \ldots + a_{n-1} \frac{dy}{dx} + a_n(x) y$$

34. **Legendre Differential Equations and Legendre Functions and Legendre's Equation**

The Legendre differential equation is

$$(1 - x^2) \frac{d^2 y}{dx^2} - 2x \frac{dy}{dx} + n(n+1) y = 0$$

Legendre's equation is $\dfrac{d}{dx}\left\{(1 - x^2) \dfrac{dy}{dx}\right\} n(n+1) y = 0$

35. First Laplace integral for $P_n(x)$ is

$$P_n(x) = \frac{1}{\pi} \int_0^\pi \left[x \pm \sqrt{x^2 - 1} \cos\phi \right]^n d\phi$$

where, n is +ve integer.

36. Bessel's differential equation is

$$x^2 \frac{d^2 y}{dx^2} + \frac{x dy}{dx} + (x^2 - n^2) y = 0$$

37. If $f(x)$ is a periodic function of x then the Fourier integral of $f(x)$ may be expressed as

$$f(x) = \frac{1}{2\pi} \int_{-\infty}^{\infty} e^{i\omega x} d\omega \int_{-\infty}^{\infty} f(t) e^{-i\omega t} dt$$

This may be expressed as

$$T(x) = \frac{1}{\sqrt{(2\pi)}} \int_{-\infty}^{\infty} e^{i\omega x} g(\omega) d\omega$$

where, $g(\omega) = \frac{1}{\sqrt{(2\pi)}} \int_{-\infty}^{\infty} f(t) e^{-i\omega t} dt.$

38. Fourier sine and cosine transforms of a function $f(t)$ are defined as

$$g_s(\omega) = \sqrt{\frac{2}{\pi}} \int_0^{\infty} f(t) \sin \omega t \, dt$$

$$g_c(\omega) = \sqrt{\frac{2}{\pi}} \int_0^{\infty} f(t) \cos \omega t \, dt$$

39. **Laurent's series** If $f(z)$ is analytic and single-valued on two concentric circles C_1 and C_2 with centre at z_0 and in the annulus between them, then $d(z)$ given by Laurent's series is

$$f(z) = \sum_{n=0}^{\infty} a_n (z - z_0) + \sum_{n=1}^{\infty} b_n (z - z_0)^{-n}$$

where,

$$a_n = \frac{1}{2\pi i} \int_C \frac{f(z') dz'}{(z' - z_0)^{n+1}}$$

$$b_n = \frac{1}{2\pi i} \int_C (z' - z_0)^{n-1} f(z') dz'$$

Again it can be expressed as

$$f(z) = \sum_{n=-\infty}^{\infty} A_n (z - z_0)^n$$

where, $A_n = \int_C \frac{f(z') dz'}{(z' - z_0)^{n+1}}.$

40. **Cauchy Residue Theorem** If a single-valued function $f(z)$ is analytic within and on a closed contour C, except for a finite number of singular points $z_1, z_2, ..., z_n$ interior to C, then

$$\int_C f(z)dz = 2\pi i \sum_{k=1}^{n} \operatorname{Res}_{z=z_k} f(z)$$

$$= 2\pi i \times (\text{sum of residues at all singularities within } C)$$

41. **Recurrence formula for $P_n(x)$** is

$$nP_n = (2n-1)xP_{n-1} - (n-1)P_{n-1}$$

42. **Recurrence formula for $J_n(x)$**

$$xJ'_n(x) = nJ_n(x) - xJ_{n+1}(x)$$

43. **Orthonormality of Bessel's function** If α and β are the roots of equation $J_n(\mu) = 0$ then the condition of orthogonality of Bessel's function over the interval (0, 1) with weight function x is

$$\int_0^1 J_n(\alpha x) J_n(\beta x) x\, dx = 0 \quad \text{for } \alpha \neq \beta$$

Condition for normality, (normalization) is

$$\int_0^1 xJ_n(\alpha x)^2 dx = \frac{1}{2} J_{n+1}^2(\alpha)$$

Both the above equations represent the conditions of orthonormality and may be written in the form of a single equation as

$$\int_0^1 J_n(\alpha x) J_n(\beta x) x\, dx = \frac{1}{2} J_{n+1}^2(\alpha) \delta_{\alpha\beta}$$

where, $\delta_{\alpha\beta}$ = Kronecker delta symbol defined as $\delta_{\alpha\beta} = 1$ for $\alpha = \beta$, and $\delta_{\alpha\beta} = 0$ for $\alpha \neq \beta$.

44. **Fourier series in the interval (0, 1): Half range series.**
Cosine series
If $f(x)$ is an even function of x then

$$f(x) = a_0 + \sum_{n=1}^{\infty} a_n \cos\frac{n\pi x}{l}$$

where, $a_0 = \dfrac{1}{l}\int_0^l f(x)\,dx$

$$a_n = \dfrac{2}{l}\int_0^l f(x)\cos\dfrac{n\pi x}{l}\,dx$$

Sine series

If $f(x)$ is odd function of x then

$$f(x) = \sum_{n=1}^{\infty} b_n \sin\dfrac{n\pi x}{l}$$

$$b_n = \dfrac{2}{l}\int_0^l f(x)\sin\dfrac{n\pi x}{l}\,dx$$

45. Complex form of Fourier series

The Fourier series for a function $f(x)$ is

$$f(x) = \sum_{n=-\infty}^{+\infty} C_n e^{inx}$$

where, $C_n = \dfrac{1}{2\pi}\int_{-\pi}^{\pi} f(x)e^{-inx}\,dx$.

46. Fourier integral

The Fourier series of periodic function $f(x)$ in the interval $(-l, +l)$ is given by

$$f(x) = a_0 + \sum_{n=1}^{\infty} a_n \cos\dfrac{n\pi x}{l} + \sum_{n=1}^{\infty} b_n \sin\dfrac{n\pi x}{l}$$

where,

$$a_0 = \dfrac{1}{2l}\int_{-l}^{l} f(x)\,dx = \dfrac{1}{2l}\int_{-l}^{l} f(t)\,dt$$

$$a_n = \dfrac{1}{l}\int_{-l}^{l} f(t)\cos\dfrac{n\pi t}{l}\,dt$$

$$b_n = \dfrac{1}{l}\int_{-l}^{l} f(t)\sin\dfrac{n\pi t}{l}\,dt$$

47. Orthogonality function for Legendre is

$$\int_{-1}^{+1} P_m(x) P_n(x)\, dx = \frac{2}{2n+1}\delta_{mn}$$

$$= 0 \text{ if } m \neq n$$

$$= \frac{2}{2n+1} \text{ if } m = n$$

48. **Hermite** Generating function of $H_n(x)$ is

$$H_n(x) = e^{x^2}(-1)^n \frac{\partial^n}{\partial x^n}(e^{-x^2})$$

49. Recurrence formula for $H_n(x)$

$$H'_n(x) = 2nH_{n-1}(x)$$

50. Orthogonal property of Hermite polynomial is

$$I_{m,n} = \int_{-\infty}^{\infty} e^{-\frac{x^2}{2}} H_m(x) H_n(x)\, dx = 2^n (n!)\sqrt{\pi}\, \delta_{m,n}$$

if $m = n$, $\delta_{mn} = 1$

if $m \neq n$, $\delta_{mn} = 0$

51. **Laguerre equation** Generating function is

$$e^{-\frac{x}{1-t}} = (1-t)\sum_{n=0}^{\infty} L_n(x)\frac{t^n}{n!}$$

52. Laguerre's differential equation is

$$x\frac{d^2y}{dx^2} + (1-x)\frac{dy}{dx} + ny = 0$$

53. Rodrigue's formula for Laguerre polynomial is

$$L_n(x) = \frac{e^x}{n!}\frac{d^n}{dx^n}(x^n e^{-x}) \to n \text{ is an integer.}$$

54. Recurrence relation for Laguerre polnomials are
   i. $(n+1)L_{n+1}(x) = (2n+1-x)L_n(x) - nL_{n-1}(x)$
   ii. $nL'_n(x) = nL_n(x) - nL_{n-1}(x)$

55. Orthogonal property of Laguerre polynomial is

$$\int_0^\infty e^{-x} L_m(x) L_n(x) dx = \delta_{m,n}$$

MULTIPLE CHOICE QUESTIONS

1. Two complex numbers Z_1 and Z_2 are given to satisfy $|Z_1|=|Z_2|$. From this it necessarily follows that
   (a) $Z_1 = Z_2$
   (b) $Z_1 = e^{i\phi} Z_2^*$ for some real ϕ
   (c) $Z_1 = Z_2^*$
   (d) $Z_1 = Z_2^{-1}$

2. A vector field $\vec{A}(\vec{r})$ is given to satisfy $\nabla \cdot \vec{A} = 0$ and $\nabla^2 \vec{A} = 0$. From these it necessarily follows that
   (a) $\vec{A} = 0$
   (b) $\nabla \times (\nabla \times \vec{A}) = 0$
   (c) $\nabla \times \vec{A} = 0$
   (d) $\vec{A}$ is a constant

3. Let $f(x)$ be an even periodic function of x which is continuous, then
   (a) it can be represented by a cosine series
   (b) it can be represented by a sine series
   (c) it can be represented by a fourier series
   (d) it needs both sine and cosine terms in its fourier expansion

4. Two Hermitian operators A and B commute (on a finite dimensional linear vector space). This implies that
   (a) every eigen vector of A is necessarily an eigen value of B
   (b) a complete set of simultaneous eigen vectors of A and B exists
   (c) every vector is necessarily an eigen vector of A and B
   (d) all eigen vectors of A are always orthogonal to those of B

5. Let A be an arbitrary $m \times n$ matrix. Let m denote the number of distinct eigen values of A and let p denote the maximum number of linearly independent eigen values of A. Then
   (a) $m = n$ for every A
   (b) $p = m$ for every A
   (c) $n \geq p \geq m$ for every A
   (d) $p = n$ for every A

6. For the differential equation $z^2 \dfrac{d^2 y}{dz^2} + 2 \dfrac{dy}{dz} + y = 0$,
   (a) $z = \infty$ is an ordinary point
   (b) $z = 0$ is a regular singular point and $z = \infty$ is an irregular singular point
   (c) $z = 0$ and $z = \infty$ both are regular singular points
   (d) $z = 0$ is an irregular singular point

7. The column element in the spherical polar coordinates is given by
   (a) $r^2 \sin\theta\, d\theta\, d\phi dr$
   (b) $r^2 \sin\theta \sin\phi d\theta\, d\phi dr$
   (c) $r^2 \sin\phi d\theta\, d\phi dr$
   (d) $r^2 d\theta\, d\phi dr$

8. Which of the following defines a conservative force $\vec{F}$?
   (a) $\dfrac{d\vec{F}}{dt} = 0$
   (b) $\nabla \cdot \vec{F} = 0$
   (c) $\nabla \times \vec{F} = 0$
   (d) $\oint \vec{F} \cdot d\vec{r} = \phi$

9. If $\dfrac{\partial L}{\partial q_n} = 0$, where L is the Lagrangian for a conservative system without constraints and q_n is a generalized coordinate then generalized momentum is
   (a) a cyclic coordinate
   (b) a constant of motion
   (c) equal to $\dfrac{d}{df}\left(\dfrac{\partial L}{\partial q_n}\right)$
   (d) undefined

10. The eigen values of the matrix $\begin{pmatrix} 0 & 1 & 0 \\ 0 & 0 & 1 \\ 1 & 0 & 0 \end{pmatrix}$ are given by λ_1, λ_2 and λ_3. Which one of the following statement is not true?
    (a) sum of the eigen value is a +ve number
    (b) $\lambda_1 \lambda_2 + \lambda_2 \lambda_3 + \lambda_3 \lambda_1 = 0$

(c) all eigen values are real

(d) product of the eigen values is 1

11. The real part of the complex function $f(x)$ is analytic in a region given by $u(x, y) = x^2 - y^2$. If the function vanishes at $z = 0$, then the imaginary part of the function is

(a) $2xy$ (b) $-2xy$ (c) $x^2 + y^2$ (d) $y^2 - x^2$

12. The residue of the function $f(z) = \dfrac{z}{(2z+1)(5-z)}$ at $z = -\dfrac{1}{2}$ is

(a) $\dfrac{1}{11}$ (b) $-\dfrac{1}{11}$ (c) $\dfrac{1}{22}$ (d) $-\dfrac{1}{22}$

13. Any two eigen vectors corresponding to two distinct eigen values of a Hermitian matrix are

(a) unitary (b) orthogonal
(c) imaginary (d) unimodular

14. For the matrix $A\begin{bmatrix} 1 & 0 \\ 0 & 2 \end{bmatrix}$, e^A is

(a) $\begin{bmatrix} e & e \\ e & e \end{bmatrix}$ (b) $\begin{bmatrix} e & 0 \\ 0 & 2 \end{bmatrix}$

(c) $\begin{bmatrix} e & 0 \\ 0 & e^2 \end{bmatrix}$ (d) $\begin{bmatrix} 0 & e \\ e^2 & 0 \end{bmatrix}$

15. The process of obtaining an orthonormal set of function from unnormalized linearly independent set of function is

(a) normalization
(b) orthogonality
(c) Schmidt ortho-normalization
(d) Laplace transform

16. If m is an integer less than n, then $\int_{-1}^{1} X^m P_n(x) dx$ is

(a) 0 (b) 1

(c) $2^{n+1}(n!)^2$ (d) $\dfrac{2^{n+1}(n!)^2}{(2n+1)!}$

17. Laplace transform of $e^{-2t} \sin 3t$ is

(a) $\dfrac{3}{S^2 - 4S + 9}$

(b) $\dfrac{3}{S^2 + 4S + 9}$

(c) $\dfrac{1}{S^2 + 4S}$

(d) $\dfrac{3}{S^2 + 4S + 13}$

18. Dirac-delta function is used when a function exists with
    (a) zero values in a very short interval
    (b) infinite values in a long interval
    (c) non-zero values in a very short interval
    (d) non-zero values in a long interval

19. If a function of two variables is a solution of Laplace's equation then the function is said to be
    (a) conjugate
    (b) harmonic
    (c) anharmonic
    (d) discontinuous

20. If C is a closed contour, then $\displaystyle\int_C dz$ is
    (a) 0
    (b) $2\pi i$
    (c) $\dfrac{1}{2\pi i}$
    (d) 2π

21. The function $f(z) = z^{-1}$ of complex variables is
    (a) not analytic
    (b) analytic at all points
    (c) conjugate
    (d) anharmonic

22. If every vector ϕ in the space can be expressed as a linear combination of $\phi_1, \phi_2, ..., \phi_n$ then the set $\phi_1, \phi_2, ..., \phi_n$ is called as
    (a) vector
    (b) basis in the vector space
    (c) scalar
    (d) field

23. If A is Hermitian matrix, then which is Hermitian for every matrix B?
    (a) $B^+ AB$
    (b) BAB
    (c) $A^+ BA$
    (d) $BA^{-1} B$

24. Any two eigen vectors of a real symmetric matrix are
    (a) orthogonal
    (b) unitary
    (c) Skew Hermitian
    (d) unimodular

25. The operator $\hat{A}$ is said to be linear if
   - (a) $\hat{A}(c_1\phi_1 + c_2\phi_2) = (c_1\hat{A}\phi_1)(c_2\hat{A}\phi_2)$
   - (b) $\hat{A}(c_1\phi_1 + c_2\phi_2) = \dfrac{(c_1\hat{A}\phi_1)}{(c_2\hat{A}\phi_2)}$
   - (c) $\hat{A}(c_1\phi_1 + c_2\phi_2) = (c_1\phi_1 + c_2\phi_2)$
   - (d) $\hat{A}(c_1\phi_1 + c_2\phi_2) = c_1\hat{A}\phi_1 + c_2\hat{A}\phi_2$

   where, c_1 and c_2 are arbitrary constants.

26. The Schwartz inequality is stated as
   - (a) $(\phi_1 \cdot \phi_1)\cdot(\phi_2 \cdot \phi_2) \geq |(\phi_1 \cdot \phi_2)|^2$
   - (b) $(\phi_1 \cdot \phi_1)\cdot(\phi_2 \cdot \phi_2) \leq |(\phi_1 \cdot \phi_2)|^2$
   - (c) $(\phi_1 \cdot \phi_1)\cdot(\phi_2 \cdot \phi_2) \geq |(\phi_1 \cdot \phi_2)|$
   - (d) $(\phi_1 \cdot \phi_1)\cdot(\phi_2 \cdot \phi_2) \leq |(\phi_1 \cdot \phi_2)|$

27. The Wronskain of the two linearly independent solutions of the Hermite differential equation is proportional to
   - (a) $\exp(x)$
   - (b) $\exp(-x)$
   - (c) $\exp(-x^2)$
   - (d) $\exp(+x^2)$

28. The non-zero solution of the Stern–Liouville problem are known as the
   - (a) eigen values of the problem
   - (b) weight function
   - (c) eigen functions of the problem
   - (d) boundary conditions

29. The value of $\int\limits_{-\infty}^{+\infty} \delta(x)dx$ is
   - (a) 1
   - (b) 0
   - (c) -1
   - (d) infinity

30. The Fourier sine transform of e^{-x} is proportional to
   - (a) $\dfrac{W}{1+W^2}$
   - (b) $\dfrac{W}{1+W}$
   - (c) $\dfrac{1}{1-W^2}$
   - (d) $\dfrac{W^2}{1+W}$

31. The argument of the quotient of two complex numbers is equal to the
    (a) sum of their arguments
    (b) difference between their arguments
    (c) product of their arguments
    (d) ratio of their arguments

32. The real and imaginary parts of a complex function $f(z) = u(x,y) + iv(x,y)$ that is analytic in a domain D have continuous second order partial derivatives and satisfy the
    (a) Poisson's equation in 3-dimensions
    (b) Bessel equation
    (c) Laplace's equation in 2-dimensions
    (d) Wave equation in 3-dimensions

33. Which of the following is a valid solution of the differential equation $\dfrac{\partial^2 \psi}{\partial x^2} + \dfrac{\partial^2 \psi}{\partial y^2} = 0$?
    (a) $x^2 - y^2$
    (b) $ax^2 - by^2$
    (c) $x^2 + y^2$
    (d) $x^4 + y^4$

34. Div curl A is equal to
    (a) div A (Curl A)
    (b) 0
    (c) div A + curl A
    (d) grad div $A - \nabla^2 A$

35. If Z_1 and Z_2 are two complex numbers then
    (a) $|Z_1 + Z_2| < |Z_1| + |Z_2|$
    (b) $|Z_1 + Z_2| = |Z_1| + |Z_2|$
    (c) $(Z_1 + Z_2)^2 > 0$
    (d) $|Z_1 + Z_2|^3 = (Z_1 + Z_2^*)(Z_1^* + Z_2)$

36. A plane central field $A rf(r)$ is said to be irrotational if
    (a) $\text{div} \cdot \vec{A} = 0$
    (b) $\text{curl} \vec{A} = 0$
    (c) $\text{grad} \vec{A} = 0$
    (d) $\text{curl curl} \vec{A} = 0$

37. A function $f(z)$ ceases to be analytic at three points
    (a) poles
    (b) zeros
    (c) singular
    (d) residues

38. A square matrix a_{ij} is said to be non-singular if
    (a) $\det a_{ij} \neq 0$
    (b) $\det a_{ij} = 0$
    (c) $\text{adj}\, a_{ij} \neq 0$
    (d) $\text{adj}\, a_{ij} = 0$

39. A function $f(x)$ is defined as half-range Fourier cosine series if
    (a) $a_o = 0$ and $a_n = 0$
    (b) $a_n = 0$ and $b_n = 0$
    (c) $a_n = 0$ and $b_2 = 0$
    (d) $b_n = 0$, a_n and a_o exists

40. The necessary and sufficient condition for the first order differential equation $P(x, y)\delta x + Q(x, y)\delta y = 0$ to be exact is
    (a) $\dfrac{\delta P}{\delta Q} = \dfrac{\delta x}{\delta y}$
    (b) $\dfrac{\delta P}{\delta y} = \dfrac{\delta Q}{\delta x}$
    (c) $\dfrac{\delta P}{\delta x} = \dfrac{\delta Q}{\delta y}$
    (d) $\dfrac{\delta P}{\delta Q} = \dfrac{\delta y}{\delta x}$

41. If a matrix $[a] = ((a)')^{-1}$ then it is called as
    (a) Hermitian
    (b) orthogonal
    (c) unitary
    (d) Skew Hermitian

42. If $\overline{A}$ is a vector the equation $\iiint_V (\nabla \cdot \overline{A}) dv = \iint_S \overline{A} \cdot d\overline{S}$ is a statement of
    (a) First Green's theorem
    (b) Stoke's theorem
    (c) Gauss's theorem
    (d) Second Green's theorem

43. If $f(z) = u(x,y) + iv(x,y)$ is analytic at z, which of the following equations does it satisfy
    (a) $\dfrac{\delta u}{\delta y} = \dfrac{-\delta v}{\delta y}$
    $\dfrac{\delta u}{\delta x} = \dfrac{\delta v}{\delta x}$
    (b) $\dfrac{\delta u}{\delta y} = \dfrac{\delta v}{\delta y}$
    $\dfrac{\delta u}{\delta x} = -\dfrac{\delta v}{\delta x}$
    (c) $\dfrac{\delta u}{\delta y} = \dfrac{\delta v}{\delta x}$
    $\dfrac{\delta u}{\delta x} = \dfrac{\delta v}{\delta y}$
    (d) $\dfrac{\delta u}{\delta y} = \dfrac{-\delta v}{\delta x}$
    $\dfrac{\delta u}{\delta x} = \dfrac{\delta v}{\delta y}$

44. If the length l of a cylinder can be measured with error a and its radius r with error b, the error in the measured area of the cylindrical surface is

(a) $2\pi(lb + ra)$
(b) $2\pi(la + rb)$
(c) $2\pi a(l + r)$
(d) $2\pi b(l + r)$

45. $F(t)$ is a symmetric periodic function of t, i.e., $F(t) = F(-t)$. Then the Fourier series $F(t)$ will be of the form

(a) $A_0 + \sum_{n=1}^{\infty} A_n \cos n\omega t + \sum_{n=1}^{\infty} B_n \sin n\omega t$

(b) $\sum_{n=1}^{\infty} A_n \cos n\omega t + \sum_{n=1}^{\infty} B_n \cos n\omega t$

(c) $A_0 + \sum_{n=1}^{\infty} A_n \cos n\omega t$

(d) $A_0 + \sum_{n=1}^{\infty} B_n \sin n\omega t$

46. The matrix $A = \begin{bmatrix} \frac{1}{2} & 0 & -\frac{\sqrt{3}}{2} \\ 0 & 1 & 0 \\ \frac{\sqrt{3}}{2} & 0 & x \end{bmatrix}$ represents a rotation for x is equal to

(a) $\frac{1}{2}$
(b) 0
(c) 1
(d) $\frac{\sqrt{3}}{2}$

47. Eigen values of a Hermitian operator are

(a) zero
(b) unimodular
(c) imaginary
(d) real

48. If $g_{\mu\nu} = 0$ for $\mu \neq \nu$ and if μ, ν, σ are unequal indices then

(a) $\Gamma_{\mu,\nu\sigma} = 0$
(b) $\Gamma_{\mu,\nu\sigma} = \frac{1}{2}$
(c) $\Gamma_{\mu,\nu\sigma} = \Gamma_{\mu\mu,\gamma}$
(d) $\Gamma_{\mu,\nu\sigma} = \frac{1}{2} \frac{\partial g_\mu}{\partial x^\mu}$

49. The value of the multiple integral $\int_0^1 dx_1 \int_0^{x_1} dx_2 \ldots \int_0^{x_{n-1}} dx_n \, x_1, x_2, \ldots, x_n$ is equal to

(a) $\dfrac{1}{n!}$ (b) $\dfrac{1}{2n!}$ (c) $\dfrac{1}{2^n}$ (d) $\dfrac{1}{2^n n!}$

50. The function $\left(1+\dfrac{1}{r}\right)z$ of a complex variable z

(a) has a simple pole at $z = 0$
(b) has a branch out from $z = 1$ to $z = \infty$ along the real axis
(c) is finite at all points inside the unit circle centred at the origin
(d) has a branch point at $z = 0$

51. The Hermitian matrices can be simultaneously diagonalized if and only if they

(a) are equal   (b) commute
(c) anticommute   (d) are square matrix

52. The eigen values of a Hermitian matrix are

(a) zero   (b) purely imaginary
(c) real   (d) unit modulus

53. A $n \times n$ real anti-symmetric matrix with odd value of n must be

(a) non-singular   (b) singular
(c) unit matrix   (d) Hermitian

54. If H is a Hermitian matrix, then e^{iH} is a

(a) Hermitian matrix   (b) Skew Hermitian matrix
(c) unitary matrix   (d) orthogonal matrix

55. The characteristic equation of the matrix $A = \begin{pmatrix} 2 & -1 & 1 \\ -1 & 2 & -1 \\ 1 & -1 & 2 \end{pmatrix}$ is

(a) $\lambda^3 - 6\lambda^2 + 9\lambda - 4 = 0$
(b) $\lambda^3 + 6\lambda^2 - 9\lambda + 4 = 0$
(c) $\lambda^3 + 6\lambda^2 + 9\lambda + 4 = 0$
(d) $\lambda^3 - 6\lambda^2 - 9\lambda - 4 = 0$

56. A real matrix is unitary if and only if it is

(a) unitary   (b) orthogonal
(c) Skew Hermitian   (d) spectral

57. A real orthogonal matrix is a special case of a
    (a) unitary matrix   (b) Hermitian matrix
    (c) Skew Hermitian matrix   (d) spectral matrix

58. The rank of a matrix remains invariant under
    (a) similarity transformation (b) elementary operations
    (c) diagonalization   (d) transitivity

59. If a vector ϕ is not normalized, a normalized vector is obtained by
    (a) $\dfrac{\phi}{\sqrt{\phi\phi}}$ (b) $\phi\sqrt{\phi\phi}$ (c) $\dfrac{\phi}{\phi\phi}$ (d) $\dfrac{\phi}{(\phi\phi)^2}$

60. If matrices A and B do not commute, then
    (a) $e^A = e^B$ (b) $e^A e^B = A + B$
    (c) $e^A e^B \neq e^{A+B}$ (d) $e^A e^B = e^{A+B}$

61. If n is an even integer, $H_n(0)$ is
    (a) 0   (b) $\dfrac{n!}{\left(\dfrac{n}{2}\right)!}$
    (c) $(-1)^n \dfrac{n!}{\left(\dfrac{n}{2}\right)!}$ (d) $(1)^n \dfrac{n!}{(n+2)!}$

62. If the Wronskain is not equal to zero, then $\sum_i \lambda_i y_i = 0$ has no solution other than
    (a) $\lambda_i = 1$ (b) $\lambda_i = -1$ (c) $\lambda_i = \infty$ (d) $\lambda_i = 0$

63. The matrix $\begin{pmatrix} 0 & 1 \\ -1 & 0 \end{pmatrix}$ is
    (a) symmetric and orthogonal
    (b) symmetric and non-orthogonal
    (c) skew symmetric and orthogonal
    (d) skew symmetric and non-orthogonal

64. The continuous complex function $f(z) = z^k = x - iy$ is differentiable
    (a) only at $z = 0$ (b) only at $z = 1$
    (c) nowhere   (d) everywhere

65. The Rodrigue's formula for Legendre polynomial is

(a) $P_n(x) = \dfrac{1}{2^n n!} \dfrac{d^n}{dx^n}(x^2 - 1)$

(b) $P_n(x) = \dfrac{d^n}{dx^n}(x^2 - 1)^n$

(c) $P_n(x) = \dfrac{d^n}{dx^n}(x^2 + 1)^n$

(d) $P_n(x) = \dfrac{1}{2^n n!} \dfrac{d^n}{dx^n}(x^2 - 1)^n$

66. The orthonormal property of Hermite polynomial is

(a) $2^n n! \pi \delta_{mn}$
(b) $2^n n! \sqrt{\pi} \delta_{mn}$
(c) $2^n \sqrt{\pi} \delta_{mn}$
(d) $2n! \sqrt{\pi} \delta_{mn}$

where, δ_{mn} is Kronecker delta.

67. Two sets of vectors one with m elements and the other with n elements ($m < n$) span the same linear vector space. If k is the dimension of the vector space then

(a) $k \leq m$
(b) $k \geq m$
(c) $k = m + n$
(d) $k = n$

68. If $g(\omega)$ is the Fourier transform of $f(t)$ the Fourier transform of $f(at)$ is

(a) $\dfrac{1}{a} g\left(\dfrac{\omega}{a}\right)$

(b) $g\left(\dfrac{-\omega}{a}\right)$

(c) $a g\left(\dfrac{\omega}{a}\right)$

(d) $e^{\pm i \omega a} g(\omega)$

69. Laplace transform of k (constant) is

(a) $\dfrac{k}{S}$ (b) S^2 (c) $\dfrac{1}{S^2}$ (d) $\dfrac{1}{kS}$

70. The vectors ϕ_1 and ϕ_2 are said to be orthogonal to each other if

(a) their sum is zero
(b) their difference is zero
(c) the scalar product $(\phi_1 \cdot \phi_2)$ is zero
(d) the vector product is zero

71. If z_0 is an ordinary point at the differential equation $u''(z) + p(z)u'(z) + q(z) = 0$, it admits in a neighbourhood of z_0, two linearly independent power series solutions of the form

(a) $u(z) = \sum_{n=0}^{\infty} (z - z_0)$
(b) $u(z) = \sum_{n=0}^{\infty} (z - z_0)^n$
(c) $u(z) = \sum_{n=0}^{\infty} C_n (z - z_0)$
(d) $u(z) = \sum_{n=0}^{\infty} C_n (z - z_0)^n$

72. Sine series for $f(x)$ when $0 \leq x \leq \pi$ is

(a) $\frac{2}{\pi} \sum_{n=1}^{\infty} \sin nx \int_0^{\pi} f(v) nv \, dv$
(b) $\sum_{n=1}^{\infty} \sin nx \int_0^{\pi} f(v) nv \, dv$
(c) $\frac{2}{\pi} \sum_{n=1}^{\infty} \sin nx$
(d) $\sum_{n=1}^{\infty} \sin nx$

73. If $g(\omega)$ is the Fourier transform of $f(t)$, the Fourier transform of $f(t \pm a)$ is (where, a is any constant.)

(a) $e^{\pm i\omega a} g(\omega)$
(b) $e^a g(\omega)$
(c) $e^{\pm i\omega a}$
(d) $\omega a g(\omega)$

74. Laplace transform of $\sinh at$ is

(a) $\frac{a}{S - a}$
(b) $\frac{a}{S^2 - a}$
(c) $\frac{a}{S - a^2}$
(d) $\frac{a}{S^2 + a^2}$

75. The value of the integral $\int_C (z^2 - 2z - 3) dz$ where, C is the circle $|z| = 2$ is

(a) 2
(b) 0
(c) $2\pi i$
(d) $\frac{1}{2\pi i}$

76. A point at which a function $f(z)$ ceases to be regular is a/an

(a) non-singular point of $f(z)$
(b) singular point of $f(z)$
(c) ordinary post
(d) pole

77. An analytic function whose only singularities in the finite plane are poles is said to be a/an
    (a) entire function   (b) meromorphic function
    (c) Laurent function   (d) Taylor's function

78. Which of the following is an analytic function of the complex variable $z = x + iy$?
    (a) $|z|$ (b) $R_e z$ (c) z^{-1} (d) $\log z$

79. Laplace transform of t is
    (a) $\dfrac{1}{S}$ (b) S^2 (c) S (d) $\dfrac{1}{S^2}$

80. The modulus of the sum of complex numbers
    (a) can never exceed the sum of their moduli
    (b) can never be less than the difference of their moduli
    (c) is the product of their moduli
    (d) is the quotient of their moduli

81. The residue theorem is useful in finding the values of
    (a) certain definite integers   (b) analytic functions
    (c) singularity   (d) harmonic functions

82. The Fourier transform of $e^{|t|}$ is
    (a) $\sqrt{\dfrac{2}{\pi}}\left(\dfrac{1}{1+\omega^2}\right)$ (b) $\left(\dfrac{1}{1+\omega^2}\right)$
    (c) $\dfrac{2}{\sqrt{\pi}}\left(\dfrac{1}{1+\omega}\right)$ (d) $\sqrt{\dfrac{2}{\pi}}$

83. Laplace transform of $\sin at$ is
    (a) $\dfrac{a}{S^2 - a^2}$ (b) $\dfrac{a}{S^2 + a^2}$
    (c) $\dfrac{-a}{S^2 + a^2}$ (d) $\dfrac{a^2}{S^2 + a^2}$

84. If $f(z)$ is analytic throughout a simply connected bounded domain D, then for every closed contour C in domain D

   (a) $f(z)d(z) = \pi$

   (b) $f(z)dz = \text{constant}$

   (c) $\int_C f(z)d(z) = 0$

   (d) $f(z) d(z) = 2\pi i$

85. The value of $H_{2n}(0)$ is

   (a) $\dfrac{2n!}{n!}$

   (b) $\dfrac{1^n 2n!}{n!}$

   (c) $(-1)^n \dfrac{2n!}{n!}$

   (d) $(-1)^n \dfrac{2n}{n}$

86. Under a unitary transformation $\hat{u}$ on a vector space a linear operator $\hat{A}$ gets transformed to the operator

   (a) $\hat{u}\hat{A}\hat{u}^{-1}$

   (b) $\hat{u}\hat{A}^{-1}\hat{u}^{+}$

   (c) $\hat{u}^{-1}\hat{A}\hat{u}$

   (d) $\hat{u}\hat{A}\hat{u}^{+}$

87. The limit of $\dfrac{1^2 + 2^2 + 3^2 + \ldots n^2}{n^3}$ as $n \to \infty$ is

   (a) 0
   (b) ∞
   (c) $\dfrac{1}{6}$
   (d) $\dfrac{1}{3}$

88. The value of $|\vec{A} \times \vec{B}|^2 + |\vec{A} \cdot \vec{B}|^2$ equals

   (a) $|2A \times 2B|^2$

   (b) $|A \times A \cdot B|^2$

   (c) $|\vec{A}|^2 |\vec{B}|$

   (d) $|\vec{A}|^2 |\vec{B}|^2$

89. If $z = x + iy$ (x and y are real), $|\exp|iz||$ is

   (a) 1

   (b) $(x^2 + y^2)^{\frac{1}{2}}$

   (c) $ix\exp(x+y)$

   (d) $\exp(-y)$

90. The solution of $\dfrac{d^2 y}{dx^2} - \dfrac{5dy}{dx} + 6y = 0$ is

   (a) $Ae^{3x} + Be^{2x}$ (b) $Ae^{-3x} + Be^{2x}$
   (c) $Ae^{3x} + Be^{-2x}$ (d) $Ae^{2x} + Be^{4x}$

91. The value of the integral $\int_0^1 \dfrac{dx}{1+x^2}$ is

   (a) $\dfrac{\pi}{2}$ (b) 0 (c) $\dfrac{\pi}{4}$ (d) $-\dfrac{\pi}{4}$

92. An operator has the eigen function sin 4x and eigen values −16. The operator is

   (a) $4\dfrac{d}{dx}$ (b) $\dfrac{d^2}{dx^2}$
   (c) $-\dfrac{d}{dx^2}$ (d) $4 + \dfrac{d}{dx} + \dfrac{d^2}{dx^2}$

93. Two vectors $\vec{A}$ and $\vec{B}$ are such that $|\vec{A} + \vec{B}| = |\vec{A} - \vec{B}|$ then the angle between $\vec{A}$ and $\vec{B}$ is

   (a) 0 (b) 90° (c) 120° (d) 180°

94. The product $\vec{A}(\vec{B} \cdot \vec{C})$ where $\vec{A}, \vec{B}, \vec{C}$ are vectors is a

   (a) scalar (b) vector
   (c) tensor (d) polar vector

95. A scalar is a tensor of rank

   (a) zero (b) one
   (c) two (d) three

96. The curl of a vector $\vec{k} = i\vec{x} + j\vec{y} + k\vec{z}$ where $\hat{i}\,\hat{j}\,\hat{k}$ are unit vectors is

   (a) 3 (b) 2 (c) 0 (d) none of the above

97. The vector $\vec{W}$ is solenoidal if

   (a) $\vec{\nabla} \cdot \vec{W} = 0$ (b) $\vec{\nabla} \times \vec{W} = 0$
   (c) $\vec{\nabla}^2 \cdot \vec{W} = 0$ (d) $\vec{\nabla} \cdot (\vec{\nabla} \times \vec{W}) = 0$

98. The components of an ordinary vector in an N-dimensional space are the components of
    (a) contravariant tensor of rank 1
    (b) covariant tensor of rank 1
    (c) mixed tensor
    (d) vector

99. If $\sqrt{\dfrac{1}{2}} = \sqrt{\pi}$, then $\sqrt{\dfrac{3}{2}}$ is equal to
    (a) $\sqrt{\dfrac{\pi}{2}}$ (b) $2\sqrt{\pi}$ (c) $\sqrt{2\pi}$ (d) $\sqrt{3\pi}$

100. The parity of $Y_{em}(\theta, \phi)$ is
    (a) $(-1)^m$ (b) $(-1)^l$ (c) $(-1)^{l+m}$ (d) 0

101. If a matrix A is satisfying the equation $A^2 = A$ then the eigen values are
    (a) 1 and -1 (b) 1 and 0
    (c) 2 and $\dfrac{1}{2}$ (d) 2 and 2

102. $|a + ib|$ is
    (a) $\sqrt{a^2 - b^2}$ (b) $\sqrt{a^2 + b^2}$
    (c) $a^2 + b^2$ (d) $a^2 - b^2$

103. The values of $i(1 - i\sqrt{3})(\sqrt{3} + i)$ is
    (a) $3 + 2i\sqrt{3}$ (b) $2 + i\sqrt{3}$
    (c) $\sqrt{3} + 2i\sqrt{3}$ (d) $2 + 2i\sqrt{3}$

104. The eigen values of $\begin{pmatrix} 0 & -1 \\ 1 & 0 \end{pmatrix}$ are
    (a) $(0, 0)$ (b) $(1, -1)$ (c) $(1, 0)$ (d) $(i, -i)$

105. If A^+ and B^+ are transposed conjugates of matrices A and B respectively, then
    (a) $(\tilde{A}^+) = A$
    (b) $(AB)^+ = B^+ A^+$
    (c) $(A+B)^+ = A^+ B^+$
    (d) $(A^+ B^+) = A$

106. If A is a Hermitian matrix, then $B^+ AB$ is
    (a) diagonal
    (b) Skew Hermitian
    (c) orthogonal
    (d) Hermitian for every matrix B

107. If the Wronskain is not equal to zero, then $\sum_i \lambda_i y_i = 0$ has no solution other than $\lambda_i = 0$. The set of functions y_i is therefore
    (a) linearly independent
    (b) linearly dependent
    (c) orthonormalized
    (d) convergents

108. If the orthonormal set of function $\phi_i(x)$ form a complete set, then any well behaved continuous function $f(x)$ may be expressed as a series of the form
    (a) $F(x) = \sum \phi_n(x)$
    (b) $F(x) = \sum_{n=0}^{\infty} a_n \phi_n(x)$
    (c) $F(x) = \sum_{n=0}^{\infty} [a_n \phi_n(x)]^2$
    (d) $F(x) = \sum a_n \phi_n(x)$

109. Two vectors ϕ_1 and ϕ_2 are said to be orthogonal to each other if the scalar product $(\phi_1 \phi_2)$ is
    (a) unity
    (b) imaginary
    (c) zero
    (d) real

110. The differential form of Hermite polynomial is
    (a) $H_n(x) = (-1)^n e^{x^2} \dfrac{d^n}{dx^n}(e^{-x^2})$
    (b) $H_n(x) = e^{x^2} \dfrac{d^n}{dx^n}(e^{-x^2})$
    (c) $H_n(x) = (-1)^n e^{x^2} \dfrac{d^n}{dx^n}(e^{x^2})$
    (d) $H_n(x) = (-1)^n e^{-x^2} \dfrac{d^n}{dx^n}(e^{x^2})$

111. If $g(\omega)$ is the Fourier transform of $f(t)$, then Fourier transform of $f(t) \cos \omega t$ is given by

(a) $\frac{1}{2} g(\omega - a)$

(b) $\frac{1}{2} g(\omega + a)$

(c) $\frac{1}{2} g(\omega - a) - \frac{1}{2} g(\omega + a)$

(d) $\frac{1}{2} g(\omega - a) + \frac{1}{2} g(\omega + a)$

112. If $y(x)$ is any differential function of x such that $y(x) = 0$ at $x = x_1, x_2, ..., x_n$ then

(a) $\delta(y(x)) = \sum_{i=1}^{n} \delta(x - x_i)$

(b) $\delta(y(x)) = \sum_{i=1}^{n} \frac{1}{\left|\frac{dy}{dx}\right|_{x=x_i}}$

(c) $\delta(y(x)) = \sum_{i=1}^{n} \frac{\delta(x - x_i)}{\left|\frac{dy}{dx}\right|_{x=x_i}}$

(d) $\delta(y(x)) = \sum_{i=1}^{n} \delta(x - x_i) \cdot \left|\frac{dy}{dx}\right|_{x=x_i}$

113. A continuous function always has

(a) a well-defined first derivative
(b) a well-defined second derivative
(c) neither of the above
(d) annulus of convergence

114. If AB is an arc of a circle $|z| = R$ having $\theta_1 << \theta << \theta_2$ and $\lim_{R \to \infty} zf(z)$ tends uniformly to b; then

(a) $\lim_{R \to \infty} \int_{AB} f(z) dz = ib(\theta_2 + \theta_1)$

(b) $\lim_{R \to \infty} \int_{AB} f(z) dz = ib(\theta_2 - \theta_1)$

(c) $\lim_{R \to \infty} \int_{AB} f(z) dz = -ib(\theta_2 - \theta_1)$

(d) $\lim_{R \to \infty} \int_{AB} f(z) dz = -ib(\theta_2 + \theta_1)$

115. If A and B are idempotent matrices, then $A + B$ will be idempotent if and only if :
    (a) $AB = BA = 0$
    (b) $\tilde{A}B = \tilde{B}A = 0$
    (c) $A^{-1}B = BA = 0$
    (d) $\tilde{A}^{-1}B = BA = 0$

116. The eigen values of a Hermitian matrix are all
    (a) zero
    (b) imaginary
    (c) real
    (d) complex number

117. The standard form of a linear homogeneous differential equation of 2nd order is $\frac{d^2y}{dx^2} + P(x)\frac{dy}{dx} + Q(x) = 0$. The solution of the equation in the neighbourhood of $x = 0$ when coefficient of series $P(x)$ and $Q(x)$ at point $x = a$ is called
    (a) ordinary point
    (b) regular point
    (c) singular point
    (d) non-singular point

118. If the Wronskain is zero over the entire range of variable $\left(\sum_i \lambda_i y_i = 0\right)$, the function y_i are
    (a) orthonormalized
    (b) linearly independent over this range
    (c) linearly dependent over this range
    (d) convergent

119. The value of $P_{(2m+1)}^{(0)}$ is
    (a) 1
    (b) 0
    (c) x
    (d) $(-1)^m \frac{2m!}{2^{2m}(m!)^2}$

120. The residue of $\frac{Z}{(Z-a)(Z-b)}$ at infinity is
    (a) +1
    (b) –1
    (c) –2
    (d) +4

121. If $g(\omega)$ is the Fourier transform of $g(t)$, the Fourier transform of $f(t \pm a)$ will be given by
    (a) $e^{+i\omega a}g(\omega)$
    (b) $e^{-i\omega a}g(\omega)$
    (c) $e^{\pm i\omega a}g(\omega)$
    (d) $g(\omega)$

122. If $f(s)$ is Laplace transform of $F(t)$, then that of $e^{at} F(t)$ will be
    (a) $f(s)$ (b) $f(s-a)$ (c) $f(s+a)$ (d) $f(sa)$

123. The equation for a pole with centre at $(-1, 1)$ and radius 3 in complex variable language is
    (a) $|z+1-i|=3$
    (b) $|z+1+i|=3$
    (c) $|z-1-i|=3$
    (d) $|z-1+i|=3$

124. A singularity of an analytic function $f(z)$ is a point where $f(z)$ ceases to be
    (a) analytic
    (b) annular
    (c) convergent
    (d) meromorphic

125. The matrix $\begin{bmatrix} 0 & -1 & 0 \\ 1 & 0 & 0 \\ 0 & 0 & 1 \end{bmatrix}$ is
    (a) orthogonal
    (b) Hermitian
    (c) symmetric
    (d) antisymmetric

126. Which of the following equations is best represented by the adjoining graph?
    (a) $xy = a$
    (b) $y = mx + a$
    (c) $y = e^x + a$
    (d) $y = ax^m$

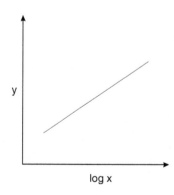

ANSWERS

1. (b)	2. (b)	3. (a)	4. (b)	5. (c)
6. (d)	7. (a)	8. (c)	9. (a)	10. (b)
11. (a)	12. (b)	13. (b)	14. (c)	15. (c)
16. (a)	17. (d)	18. (c)	19. (b)	20. (c)
21. (b)	22. (b)	23. (c)	24. (a)	25. (a)
26. (a)	27. (c)	28. (a)	29. (a)	30. (a)
31. (b)	32. (c)	33. (a)	34. (b)	35. (a, b)
36. (b)	37. (c)	38. (a)	39. (d)	40. (c)
41. (b)	42. (c)	43. (d)	44. (b)	45. (c)
46. (a)	47. (d)	48. (a)	49. (d)	50. (a)
51. (b)	52. (a)	53. (b)	54. (c)	55. (a)
56. (b)	57. (a)	58. (a)	59. (d)	60. (a)
61. (c)	62. (d)	63. (c)	64. (c)	65. (b)
66. (c)	67. (c)	68. (a)	69. (a)	70. (c)
71. (d)	72. (d)	73. (a)	74. (d)	75. (b)
76. (b)	77. (b)	78. (c)	79. (a)	80. (b)
81. (a)	82. (a)	83. (b)	84. (b)	85. (c)
86. (d)	87. (d)	88. (d)	89. (d)	90. (a)
91. (d)	92. (b)	93. (b)	94. (b)	95. (a)
96. (c)	97. (a)	98. (d)	99. (a)	100. (c)
101. (a)	102. (b)	103. (d)	104. (b)	105. (b)
106. (d)	107. (a)	108. (b)	109. (c)	110. (a)
111. (d)	112. (c)	113. (c)	114. (b)	115. (a)
116. (c)	117. (a)	118. (c)	119. (c)	120. (b)
121. (c)	122. (b)	123. (a)	124. (a)	125. (d)
126. (d)				

4

SOLID STATE PHYSICS

FORMULAE

1. Lattice translational operator is
$$T = ua + vb + wc$$
   where,
   u, v, w = integers
   a, b, c = crystal axes

2. The axis vectors of the reciprocal lattice is
$$A = 2\pi \frac{b \times c}{a \cdot b \times c}, B = 2\pi \frac{c \times a}{a \cdot b \times c}, C = 2\pi \frac{a \times b}{a \cdot b \times c}$$
   where,
   a, b, c = primitive vectors of the crystal lattice
   A, B, C = primitive vectors of reciprocal lattice

3. The reciprocal lattice vector G can be given as
$$G = hA + kB + lE$$
   where, h, k, l = integers.

4. Various statements of Bragg's condition are
$$2d \sin \theta = n\lambda, \Delta k = G \text{ and } 2k.G = G^2$$
   where,
   d = the distance $\frac{b}{n}$ on the lattice planes
   Δk = scattering vector

5. Laue conditions can be given as
$$a \cdot \Delta k = 2\pi h; \; b \cdot \Delta k = 2\pi k; \; c \cdot \Delta k = 2\pi l$$

6. Any function invariant under a lattice translation may be expanded in a Fourier series of the form
$$n(r) = \sum_G n_G \exp(iG \cdot r)$$
where, $n(r)$ = total concentration of electron at r due to all atoms in the cell.

7. The electrostatic energy of a structure of $2N$ ions of charge $\pm q$ is
$$u = -N\alpha \frac{q^2}{R} = -N\Sigma + \frac{q^2}{r_{ij}}$$
where,
α = Madelung constant
R = distance $\dfrac{b}{n}$ nearest neighbours

8. When a phonon of wave vector K is created by inelastic scattering of a phonon or neutron from wave vector k to k', the wave vector selection rules that govern the process is
$$k = k' + K + G$$
where, G is the reciprocal lattice vector.

9. If there are p atoms in the primitive cell, the phonon dispersion relation will have 3 acoustical phonon branches and $3p-3$ optical phonon branches.

10. The lattice heat capacity at constant volume is defined as
$$C_v = T\Sigma \left(\frac{\partial S}{\partial T}\right)_v = \left(\frac{\partial U}{\partial T}\right)_v$$
where,
S = entropy
U = energy
T = temperature

11. If there are N primitive cells in the specimen, the total number of acoustic phonon modes in N, and the cut-off frequency ω_D is determined by

$$\omega_D^3 = \frac{6\pi^2 v^3 N}{V}$$

where,
v = velocity of sound
V = volume of the specimen

12. The Debye temperature θ in terms of ω_D is given as

$$\theta = \frac{\hbar}{K_B} v \left[\frac{6\pi^2 N}{V}\right]^{\frac{1}{3}}$$

13. Debye's T^3 law can be written as

$$C_v = \frac{12\pi^4}{5} NK_B \left(\frac{T}{\theta}\right)^3$$

$$= 234 NK_B \left(\frac{T}{\theta}\right)^3$$

where,
N = number of atoms in the specimen
θ = Debye's temperature

14. From the kinetic theory of gases, the following expression for thermal conductivity can be written as

$$k = \frac{1}{3} Cvl$$

where,
C = heat capacity per unit volume
v = average particle velocity
l = mean free path of the particle $\frac{b}{n}$ collision

15. Density of states can be written as

$$D_\varepsilon = \frac{dN}{d\varepsilon} = \frac{V}{2\pi^2}\left(\frac{2m}{\hbar^2}\right) \varepsilon^{\frac{1}{2}} = \frac{3N}{2\varepsilon}$$

where,

ε = orbital energy
N = No. of free electrons

16. Electrical conductivity is

$$\sigma = \frac{ne^2\tau}{m}$$

Electrical resistivity is

$$\rho = \frac{m}{ne^2\tau}$$

where,

n = No. of electrons
τ = collision time
m = mass of the electron

17. Hall coefficient is given as

$$R_H = \frac{-1}{ne}$$

18. **Bloch function** The solution of Schrodinger equation for periodic potential must be of the form

$$\psi_k(r) = u_k(r)\exp(ik \cdot r)$$

19. The Peltier coefficient is related to the thermoelectric power by the relation

$$\pi = QT$$

where, T is the temperature.

20. The effective mass m^* of the electron at the wave vector k is given by

$$\left(\frac{1}{m^*}\right)_{\mu\upsilon} = \frac{1}{\hbar^2}\frac{\partial^2 \varepsilon}{\partial K_\mu \partial K_\upsilon}$$

where, μ and υ are cartesian coordinates.

21. If the electron is missing from the state of the wave vector K_e then the wave vector of the hole is

$$K_h = -K_e$$

22. The periodicity in the de Haas–van Alphen effect measures the external cross-sectional area S in the K-space of the Fermi surface, the cross-section being taken perpendicular to B, and is given by

$$\Delta\left(\frac{1}{B}\right) = \frac{2\pi e}{\hbar c S}$$

where, B is the magnetic field.

23. Definition for dieletric function is

$$D = \varepsilon_0 E + P$$

where,
E = electric field
P = polarization
ε_0 = dielectric constant

24. The plasma frequency is defined by

$$w_P^2 = \frac{n_e^2}{\varepsilon_0 m}$$

where, n is the electron concentration.

25. The screened potential energy of electron–proton pair is

$$U(R) = \frac{-e^2}{r} e^{-K_s r}$$

26. The reflectance at normal incidence is

$$R = \frac{(n-1)^2 + K^2}{(n+1)^2 + K^2}$$

where,
n = refractive index
K = extinction coefficient

27. The London equation

$$\text{Curl } j = \frac{-C}{4\pi\lambda_L^2} B$$

where,
λ_L = London penetration length
C = speed of light
B = magnetic field applied

$$\lambda_L = \left[\frac{mc^2}{4\pi n e^2}\right]^{\frac{1}{2}}$$

28. For, type-II superconductors have $\xi < \lambda$, the critical fields are connected by the relation

$$H_{C_1} \approx \left(\frac{\xi}{\lambda}\right) H_C \text{ and}$$

$$H_{C_2} \approx \left(\frac{\lambda}{\xi}\right) H_C$$

where,

ξ = coherence length

λ = penetration depth

29. The polarization P can be written as

$$P = \varepsilon_0 \chi E$$

where,

χ = dielectric susceptibility

E = applied field

30. The dielectric constant ε of an isotropic medium is defined in terms of the macroscopic field E as

$$\varepsilon = \frac{E + 4\pi P}{E} = 1 + 4\pi\chi \qquad \text{(in CGS)}$$

$$\varepsilon = \frac{\varepsilon_0 E + P}{\varepsilon_0 E} = 1 + \chi \qquad \text{(in SI)}$$

31. Susceptibility related to dielectric constant

$$\chi = \frac{P}{\varepsilon_0 E} = \varepsilon - 1$$

where,

P = polarization

E = applied field

χ = susceptibility

ε = dielectric constant

32. Clausius–Mosotti relation is
$$\frac{\varepsilon - 1}{\varepsilon + 2} = \frac{1}{3\varepsilon_0} \Sigma N_j \alpha_j$$
where,

N_j = concentration of atoms
α_j = polarizability of atoms

33. The piezoelectric equations are given by
$$p = Zd + E\chi$$
$$e = Zs + Ed$$
where,

E = applied field to the crystal
p = polarization
χ = dielectric susceptibility
d = piezoelectric strain constant
s = elastic compliance constant
e and z have the usual meaning.

34. The magnetic susceptibility per unit volume is defined as
$$\chi = \frac{M}{B}$$
where,

M = magnetization
B = magnetic field intensity

If χ is negative, the substance is diamagnetic.
If χ is positive, the substance is paramagnetic.

35. The magnetization of free electron gas is
$$M = \frac{N \mu_B^2}{K_B T_F} B$$
where,

N = No. of atoms per unit volume
μ_B = Bohr magneton
T_F = Fermi temperature

36. The diamagnetic susceptibility of N atoms of atomic number Z is
$$\chi = \frac{-Ze^2 N <r^2>}{6mc^2}$$
where, $<r^2>$ is the mean square atomic radius.

37. Atoms with permanent magnetic moment μ have paramagnetic susceptibility

$$\chi = N \frac{\mu^2}{3K_B T}$$

for $\mu B \ll K_B T$

38. The susceptibility of a ferromagnet above Curie temperature has the form

$$\chi = \frac{C}{T - T_C}$$

where, C = Curie constant.

39. Cure–Weiss law is

$$T_c = C\lambda$$

where, λ is a constant independent of temperature.

40. In an antiferromagnet, the susceptibility above Neel temperature has the form

$$\chi = \frac{2C}{T + \theta}$$

MULTIPLE CHOICE QUESTIONS

1. Which of the following metals is not ferromagnetic?
   (a) Fe (b) Ni (c) Co (d) Mn

2. What is the significance of Laue's experiment?
   (a) X-rays are transverse em waves
   (b) atoms in crystal have a regular arrangement
   (c) both (a) and (b)
   (d) X-rays have a particle nature

3. The Fermi level represents
   (a) the top edge of the highest completely filled band
   (b) the level midway between the upper edge of the valence-band and the lower edge of the CB

(c) the level of donor impurity

(d) the level up to which all energy states are occupied at 0 K.

4. If the potential energy of a linear lattice is given by $V = ax^2$, where, x is the displacement of an atom, the thermal expansion is
   (a) infinity
   (b) finite
   (c) zero
   (d) negative

5. The Hamiltonian of a non-relativistic free electron in a magnetic field B is given by $H = \left(\frac{1}{2}m\right)\left(P + \left(\frac{e}{c}\right)A\right)^2 - \mu B$. The existence of polyparamagnetism can be proved with
   (a) the A term only
   (b) the B term only
   (c) the A and B terms
   (d) with the addition of another term of H

6. The bonding in graphite is
   (a) metallic
   (b) covalent
   (c) van der Waals
   (d) combination (b) and (c)

7. The number of independent elastic constants for a cubic crystal is
   (a) 2
   (b) 3
   (c) 4
   (d) 6

8. As the interatomic distance in a solid increases, the width of an allowed energy band
   (a) increases
   (b) decreases
   (c) remains unaltered
   (d) first increases and then decreases

9. Ionic crystals are very good insulators at room temperature. They become conducting at higher temperature. This is due to
   (a) thermal expansion of the ionic solid
   (b) generation of more free electrons
   (c) transition of large number of electrons from VB to CB.
   (d) movement of large number of ions in vacancies.

10. If there are 10^{22} quantum states/m³ at $E = E_f$ at a certain non-zero temperature, then the number of electrons/m³ to be found at this energy is
    (a) 5×10^{21}
    (b) 10^{22}
    (c) 2×10^{22}
    (d) 5×10^{22}

11. A typical semiconductor has a band gap of 1 eV. It is doped with an impurity. Its effective mass of $\bar{e}_s$ in the semiconductor is $m/10$ (m = mass of the electron and dielectric constant is 16). If the ionization energy of hydrogen is 13.6 eV, the ionization energy of the impurity atoms will be

(a) 13.6 eV (b) 5.3 meV (c) 0.53 meV (d) 0.053 meV

12. Given that r is the radius of the atoms and a is the lattice constant of a solid having a cubic structure, which one of the following relation is true for a body centred cubic structure

(a) $a = 2r$ (b) $a = 2r\sqrt{2}$ (c) $a = 2r\sqrt{3}$ (d) $a = 4r\sqrt{3}$

13. Across the interface of two linear media with dielectric constants $\varepsilon_1, \varepsilon_2$ and permeabilities μ_1 and μ_2, which one of the following holds true?

(a) $\varepsilon_1 E_{1\perp} = \varepsilon_2 E_{2\perp}$ and $\dfrac{1}{\mu_1} B_{1\parallel} = \dfrac{1}{\mu_2} B_{2\parallel}$

(b) $E_{1\perp} = E_{2\perp}$ and $\mu_1 B_{1\perp} = \mu_2 B_{2\perp}$

(c) $E_{1\parallel} = E_{2\parallel}$ and $B_{1\parallel} = B_{2\parallel}$

(d) $\varepsilon_1 E_{1\parallel} = \varepsilon_2 E_{2\parallel}$ and $B_{1\perp} = B_{2\perp}$

14. The total energy per particle of a collection of free fermions is 3 eV. The Fermi energy of the system is

(a) 1.8 eV (b) 3 eV (c) 4 eV (d) 5 eV

15. Which of the following is not a property of conventional superconductors?

(a) the superconductors are perfect diamagnetics.

(b) superconductivity can be destroyed by application of a magnetic field.

(c) the specific heat of superconductors decreases exponentially with decrease in the temperature.

(d) the energy spectrum of a superconductor shows a band gap of the order of 1 eV.

16. A lattice is characterized by the following primitive vectors (in Angstroms) $\vec{a} = 2(\vec{i}+\vec{j}); \vec{b} = 2(\vec{j}+\vec{k}); \vec{c} = 2(\vec{k}+\vec{i})$. The reciprocal lattice corresponding to the above is

(a) bcc lattice with cube edge π Å

(b) bcc lattice with cube edge 2π Å

(c) fcc lattice with cube edge π Å

(d) fcc lattice with cube edge 2π Å

17. On lightly doping an intrinsic semiconductor with donor impurities, the $\bar{e}$ carrier concentration in the conduction band is found to decrease from n to n_f. If μ_n and μ_p are the mobilities of $\bar{e}s$ and holes respectively, then conductivity of the doped semiconductor will be

(a) $n_e(f\mu_n + \mu_{p/f})$
(b) $n_e(f\mu_n + \mu_p)$
(c) $n_e f \mu_n$
(d) $n_e f(\mu_n + \mu_p)$

18. The total energy of an ionic solid is given by an expression $V = \dfrac{\alpha e^2}{4\pi\varepsilon_0 r} + \dfrac{B}{r^9}$, where, α is Madelung constant and r is the distance between the nearest neighbours in the crystal. If r_0 is the equilibrium separation between the nearest neighbours, the constant B is given by

(a) $\dfrac{\alpha e^2 r_0^8}{36\pi\varepsilon_0}$
(b) $\dfrac{2\alpha e^2 r_0^8}{9\pi\varepsilon_0}$
(c) $\dfrac{\alpha e^2 r_0^{10}}{36\pi\varepsilon_0}$
(d) $-\dfrac{\alpha e^2 r_0^{10}}{36\pi\varepsilon_0}$

19. The packing in the hcp structure is

(a) ABCABC
(b) ABABAB
(c) ACACAC
(d) ACBACB

20. First Brilliouin's zone for bcc lattice is

(a) rhombic dodecahedron
(b) truncated octahedron
(c) rhombohedron
(d) tetrahedron

21. The strongest bond is

(a) covalent bond
(b) ionic bond
(c) metallic bond
(c) tetrahedron bond

22. In the phonon dispersion relation, if there are 3 acoustical phonon branches and 3 optical phonon branches then the number of atoms in the primitive cell is

(a) 1
(b) 2
(c) 3
(d) 4

23. Lorentz number L is defined as

   (a) $L = \dfrac{K}{\sigma}$ (b) $L = \dfrac{\sigma}{K}$ (c) $\dfrac{K}{\sigma T}$ (d) $L = K\sigma T$

24. In a semiconductor the fermi level lies in the
    (a) middle of the conduction band
    (b) middle of the band gap
    (c) middle of the VB
    (d) top of the VB

25. Bohr magneton μ_B in SI units is

   (a) $\dfrac{e\hbar}{2m}$ (b) $\dfrac{e\hbar}{2mc}$ (c) $\dfrac{e\hbar}{mc}$ (d) $\dfrac{em}{\hbar c}$

26. The energy which directs the magnetization in a ferromagnetic crystal along certain crystallographic axes is called
    (a) fermi energy   (b) internal energy
    (c) anisotropic energy   (d) isotropic energy

27. The distance within which the superconducting electron concentration cannot change drastically in a spatially varying magnetic field is
    (a) penetration depth   (b) skin depth
    (c) electron mean free path (d) coherence length

28. A superconductor behaves like a
    (a) diamagnet   (b) paramagnet
    (c) ferromagnet   (d) ferrimagnet

29. The packing in the fcc structure is
    (a) ABCABC   (b) ABABAB
    (c) ACACAC   (d) ACBACB

30. The reciprocal lattice to the hcp lattice is
    (a) fcc lattice   (b) bcc lattice
    (c) hcp lattice   (d) sc lattice

31. The binding of molecular hydrogen is a simple example of the
    (a) ionic bond   (b) covalent bond
    (c) hydrogen bond   (d) metallic bond

32. If there are p atoms in the primitive cell, the phonon, dispersion relation will have
    (a) 2 acoustical branches and $3p-2$ optical branches
    (b) 3 acoustical branches and $2p-3$ optical branches
    (c) 3 acoustical branches and $3p-3$ optical branches
    (d) 2 acoustical branches and $2p-2$ optical branches

33. Fermi temperature of a system
    (a) is the temperature of the system
    (b) is always zero
    (c) has nothing to do with the temperature of the system
    (d) is $\dfrac{E_F}{T}$

34. Electron wave function in a solid should obey
    (a) Bloch's theorem
    (b) Law of mass action
    (c) the relation $\psi_k(r) = \psi_k^*(r)$
    (d) the 2nd law of Newton

35. According to Hund's rule the J value of $Cr^{3+}(4f^1 5s^2 p^6)$ is
    (a) $\dfrac{3}{2}$ (b) $\dfrac{7}{2}$ (c) $\dfrac{5}{2}$ (d) $\dfrac{1}{2}$

36. In an antiferromagnet the susceptibility above the Neel temperature has the form
    (a) $\chi = \dfrac{2C}{(T+T_N)}$ (b) $\chi = \dfrac{2C}{(T-T_N)}$
    (c) $\chi = \dfrac{2C}{(T-T_C)}$ (d) $\chi = \dfrac{2C}{(T+T_N)}$

37. The spin of cooper pair is
    (a) 1   (b) 0   (c) $\dfrac{1}{2}$ (d) 2

38. Length connected with Meissner effect is
    (a) wavelength of the cooper pair
    (b) penetration depth
    (c) coherence length
    (d) normal electron mean free path

39. The atomic packing factor (APF) for the bcc metal structure is
    (a) 0.68   (b) 0.74   (c) 0.50   (d) 1.00

40. At atmospheric pressure, a material of unknown composition shows three phases in equilibrium at 987 K. What is the minimum number of components in the system?
    (a) 4   (b) 3   (c) 2   (d) 1

41. The coordination number of the zinc blend's structure is
    (a) 4   (b) 6   (c) 8   (d) 10

42. When diffusion is activated, the dependence of the diffusion coefficient on temperature is
    (a) $\exp\left(-\dfrac{E}{KT}\right)$
    (b) $\exp\left(\dfrac{E}{KT}\right)$
    (c) T
    (d) $\dfrac{1}{T}$

43. The Burger's vector of an imperfect dislocation is
    (a) always less than one lattice spacing
    (b) always more than one lattice spacing
    (c) an integral number of lattice spacing
    (d) not an integral number of lattice spacing

44. The elastic strain obtained on applying stress to a material is
    (a) time-dependent
    (b) partially permanent
    (c) instantaneous
    (d) inversely proportional

45. The eutectoid mixture in steel is
    (a) a mixture of ferrite and cementite
    (b) a mixture of ferrite and austentite
    (c) a mixture of austentite and cementite
    (d) coiled lead brite

46. Silicon carbide is
    (a) ionically bonded
    (b) produced by the controlled denitrification of glass
    (c) produced by reaction bonding
    (d) soft for a ceramic

47. The molecular weight of vinyl chloride is 62.5. Then the molecular weight of polyvinyl chloride with degree of polymerization 20,000 is
    (a) 320 (b) 3.1×10^{-3} (c) 1.25×10^6 (d) 62.5

48. Magnet recording tape is most commonly made from
    (a) small particles of iron
    (b) silicon–iron
    (c) ferric oxide
    (d) metallic glass

49. The crystal structure of diamond is
    (a) fcc with two atom basis of 10001 and $\dfrac{a}{4(\vec{i}+\vec{j}+\vec{k})}$
    (b) simple cubic with two atoms basis of 1000 and $\dfrac{a}{2(\vec{i}+\vec{j}+\vec{k})}$
    (c) fcc with two atom basis of 10001 and $\dfrac{a}{2(\vec{i}+\vec{j}+\vec{k})}$
    (d) bcc with an atom basis

50. The origin of van der Waals interaction in a molecular crystal is
    (a) nuclear
    (b) magnetic
    (c) ionic
    (d) fluctuating dipolar

51. In the original BCS model of superconductivity, the dependence of T_c on isotope mass is
    (a) $T_c \propto M'$
    (b) $T_c \propto M$
    (c) $T_c \propto M^{-\frac{1}{2}}$
    (d) $T_c \propto M^{\frac{1}{2}}$

52. At low temperatures the specific heat of insulating crystals varies as
    (a) AT^3
    (b) $BT + CT^3$
    (c) $D \exp\dfrac{E}{T}$
    (d) will not change with temperature

53. The band gap at 300 K of diamond cubic crystals of C, Si, Ge and Sn is 5, 1, 12, 0.69 and 0.1 eV respectively. Identify the crystal which will have the highest electrical conductivity at 300 K.
    (a) Ge (b) Sn (c) C (d) Si

54. Which of the following expressions is correct for a dielectric?

(a) $\oint \vec{D} \cdot da = (Q_{free})$ enclosed

(b) $\oint \vec{E} \cdot da = \dfrac{1}{\varepsilon_0}(Q_{free})$ enclosed

(c) $\oint \vec{D} \cdot da = \dfrac{1}{\varepsilon_0}(Q_{free})$ enclosed

(d) $\oint \vec{D} \cdot da = (Q_{total})$ enclosed

55. Magnetic field of an infinitely long ideal solenoid of radius R carrying current i
    (a) increases radially inside and is zero outside the solenoid.
    (b) is constant inside and zero outside the solenoid.
    (c) is constant inside and decays as $\dfrac{e}{r}$ outside the solenoid.
    (d) is constant inside and decays as $\exp\left(\dfrac{1}{r}\right)$ outside.

56. In the Meissner state, the magnetic susceptibility of a superconductor in SI units is
    (a) $-\dfrac{1}{4\pi}$ (b) $\dfrac{1}{4\pi}$ (c) -1 (d) $\dfrac{1}{\mu_0}$

57. The 0.20 X-ray powder diffraction pattern of ionic crystal KCl and KBr both of which form fcc structure will have the following properties.
    (a) Same number of diffraction lines with equal intensities.
    (b) Pattern of KBr will have more lines.
    (c) Pattern of KCl will have more lines.
    (d) Number of lines will be same but their intensities will be different.

58. It is easier to excite the optical phonon branch in RbBr crystals with electromagnetic radiation because
    (a) Rb and Br ions have same number of electrons.
    (b) Rb and Br ions are oppositely charged.
    (c) Rb and Br ions have nearly the same atomic weight.
    (d) RbBr is an insulator.

59. When the *em* waves interact with a metal with active acoustic and optical phonon, the *em* wave will not propagate if its frequency lies between
    (a) ω_L and ω_p
    (b) ω_L and zero
    (c) ω_L and ω
    (d) ω_L and ω_T

60. Electric susceptibility is
    (a) independent of temperature
    (b) linearly proportional to temperature
    (c) proportional to square of the temperature
    (d) complex function of the temperature

61. The range of wave vector K that corresponds to the first Brillouin zone in metal is
    (a) $\left(-\dfrac{a}{\pi} \text{ to } +\dfrac{a}{\pi}\right)$
    (b) $\left(-\dfrac{\pi}{2a} \text{ to } +\dfrac{\pi}{2a}\right)$
    (c) $\left(-\dfrac{\pi}{a} \text{ to } +\dfrac{\pi}{a}\right)$
    (d) $\left(-\dfrac{a}{2\pi} \text{ to } +\dfrac{a}{2\pi}\right)$

62. The piezo-electric transducer can be used to measure
    (a) static variables
    (b) dynamic variables
    (c) eddy variables
    (d) temperature parameter

63. The four types of Bravais lattices—primitive, body centred, base centred and face centred—exist in only one crystal system. Identify the system.
    (a) triagonal
    (b) tetragonal
    (c) orthorhombic
    (d) cubic

64. In an ionic crystal like NaCl, the dielectric constant has contributions from
    (a) electronic polarizability only
    (b) electronic and ionic polarizability
    (c) ionic polarizability only
    (d) orientational polarizability only

65. The width of an energy band in solid
    (a) increases with interatomic separation
    (b) decreases with interatomic separation
    (c) is independent of interatomic separation
    (d) (a) or (b) depending on the structure of solid

66. The Hall probe used to measure magnetic fields contains
    (a) only p-type semiconductor
    (b) only n-type semiconductor
    (c) either (a) or (b)
    (d) metal sample

67. In the hcp structure
    (a) equidistant spheres are arranged in a regular array to maximize the interstitial volume
    (b) the C/a ratio is nearly 1.633 in the ideal case
    (c) the coordination number is less than 10
    (d) the fraction of the total volume filled by the sphere is different from fcc structure

68. If the energy of an X-ray photon is 10 keV. Its wavelength is roughly
    (a) 12.4Å   (b) 1.24Å   (c) 0.62Å   (d) 0.31Å

69. The fundamental relation in the theory of elastic scattering in a periodic lattice is
    (a) $\vec{K} \cdot \vec{G} = |\vec{G}|$
    (b) $\vec{K} \cdot \vec{G} = G^2$
    (c) $2\vec{K} \cdot \vec{G} = G^2$
    (d) $\vec{K} \times \vec{G} = \vec{G}$

70. Magnons are elementary excitation in a
    (a) ferromagnetic or antiferromagnetic substance
    (b) diamagnetic substance only
    (c) paramagnetic substance only
    (d) ferrimagnetic substance only

71. Due to the free electrons, metals have a paramagnetic susceptibility which
    (a) increases slowly with temperature
    (b) decreases slowly with temperature
    (c) increases very fast as it approaches a critical temperature
    (d) is independent of temperature

72. In crystals, symmetry axes exist pertaining to rotation by an angle $\frac{2\pi}{n}$ where, n has definite integral values. Which of the following values of n is permitted?
    (a) 5   (b) 6   (c) 7   (d) 8

73. In the NaCl structure, if the d atoms are removed, we get a lattice of Na atoms which is
    (a) hcp   (b) sc   (c) bcc   (d) fcc

74. The Wiedeman–Franz ratio relates
    (a) ionic conductivity and ionic diffusion
    (b) thermal conductivity and thermal expansion
    (c) electrical and thermal conductivities of metal
    (d) none of the above

75. Identify the physical property of crystals which would be absent if the lattice vibrations were harmonic
    (a) thermal expansion   (b) specific heat
    (c) compressibility   (d) lattice energy

76. The dielectric constant is nearly equal to
    (a) n (b) n^2 (c) $\frac{1}{n}$ (d) $n^2 - 1$

77. The thickness of Bloch walls in ferromagnetic materials is of the order of:
    (a) fraction of Å   (b) 10 Å
    (c) 100 Å   (d) 1000 Å

78. As interatomic spacing increases, an energy band
    (a) narrows
    (b) broadens
    (c) remains unchanged in width
    (d) overlaps with next band

79. The equation $\psi = \dfrac{C}{(T + \theta)}$ describes the behaviour of
    (a) diamagnetic materials
    (b) paramagnetic materials
    (c) ferromagnetic materials
    (d) antiferromagnetic materials

80. In what respect are diamond and germanium dissimilar?
    (a) crystal structure   (b) electrical conduction
    (c) nature of bonding   (d) chemical group

81. The magnet susceptibility of a degenerated paramagnetic electron gas varies at low temperature as
    (a) $\dfrac{1}{T}$
    (b) $\dfrac{1}{T-T_i}$
    (c) $\dfrac{1}{T^2}$
    (d) independent of T

82. A free electron cannot absorb a photon because
    (a) energy and momentum conservation cannot be satisfied simultaneously
    (b) photon interacts with electron only in the presence of a nucleus
    (c) the statement of the question is wrong
    (d) a free $\bar{e}$ has too much zero point momentum

83. The reciprocal lattice of a simple cubic lattice is
    (a) bcc lattice
    (b) fcc lattice
    (c) simple cubic lattice
    (d) orthorhombic lattice

84. In a silicon crystal the binding between atoms can be characterized as
    (a) ionic
    (b) van der Waals type
    (c) metallic
    (d) covalent

85. If σ is the electrical conductivity of a metal and K its thermal conductivity, the function of σ and K which is constant at a given temperature is
    (a) $K\sigma^2$
    (b) $K^2\sigma$
    (c) $\dfrac{\sigma}{K}$
    (d) $K\sigma$

86. In a ferromagnetic material, the spins are aligned parallel
    (a) at all temperatures
    (b) below a critical temperature
    (c) above a critical temperature
    (d) at a critical temperature

87. The units of Madelung constant are
    (a) cm^{-1}
    (b) cal/mole
    (c) coulomb/cm
    (d) no units

88. If σ is the conductivity of a semiconductor at temperature T, the $\log \sigma$ Vs T^{-1} plot is
    (a) linear
    (b) linear and parallel to T^{-1} axis
    (c) exponential
    (d) parabolic

89. The crystal used in the scintillation counter is
    (a) NaI
    (b) NaI (Tl)
    (c) NaCl
    (d) KCl (Tl)

90. The energy gap width for silicon (in eV) is
    (a) 0.1
    (b) 10.0
    (c) 0.011
    (d) 1.1

91. If $a = b = c$ and $\alpha = \beta = \gamma = 90°$ the crystal class is
    (a) tetragonal
    (b) triagonal
    (c) triclinic
    (d) orthorhombic
    (e) cubic

92. The transition from a normal metal to a type I superconductor
    (a) is a second-order phase transition
    (b) occurs via the formation of bosons by pairing of electrons
    (c) involves no latent heat
    (d) lead to a perfect diamagnetism

93. The Fermi surface of metal can be studied using
    (a) neutron scattering
    (b) positron annihilation
    (c) the de Hass–Van Alphen effect
    (d) magnetoresistance

94. When a mirror plane is combined with simultaneous translation operation in a crystal one obtains
    (a) screw axis
    (b) glide plane
    (c) inversion
    (d) Bravais lattice

95. The atomic packing factor for the fcc metal structure is
    (a) 0.74
    (b) 0.68
    (c) 0.50
    (d) 1.00

96. The types of lattice formed in a crystal is physically determined by the
    (a) plane patterns
    (b) point symmetry of the basis
    (c) space group
    (d) rhombus lattice

97. The atomic packing factor for hcp metal structure is
    (a) 0.74
    (b) 1.00
    (c) 0.50
    (d) 0.68
98. The volume of a unit cell of reciprocal lattice is
    (a) inversely proportional
    (b) directly proportional
    (c) equal
    (d) double
99. The diffusivity can be expressed as
    (a) $D = e^{-\theta/KT}$
    (b) $D = D_0 e^{-\theta/KT}$
    (c) $D = D_0 e^{\theta/KT}$
    (d) $D = e^{\theta/KT}$
100. A diffraction pattern of a crystal is a map of the
    (a) real crystal structure
    (b) Miller indices
    (c) reciprocal lattice of the crystal
    (d) translation vectors
101. Ion vacancies are called as
    (a) Frenkel defect
    (b) interstitial defect
    (c) Schottky defect
    (d) edge dislocation
102. The hydrogen bond is essentially
    (a) metallic
    (b) dispersive
    (c) covalent
    (d) ionic
103. For a transformation that occurs on cooling, the Gibb's free energy per unit volume is
    (a) positive
    (b) negative
    (c) zero
    (d) very high value
104. The number of slip systems axiable in nickel are
    (a) 6
    (b) 3
    (c) 12
    (d) 24
105. Covalent crystals are quite
    (a) hard but brittle
    (b) hard but ductile
    (c) soft but brittle
    (d) soft but ductile

106. At the melting point, the free energy of the melt is
    (a) zero
    (b) negative
    (c) positive
    (d) infinity

107. True strain is expressed as
    (a) $\varepsilon' = (1+\varepsilon)$
    (b) $\varepsilon' = \log_{10}(1+\varepsilon)$
    (c) $\varepsilon' = \log_e (1+\varepsilon)$
    (d) $\varepsilon' = \log_e (1-\varepsilon)$

108. For an isotropic cubic crystal, the number of moduli of elasticity is/are
    (a) one
    (b) three
    (c) two
    (d) four

109. Heteropolar molecules like HF and H_2O possesses relatively
    (a) very large dipole moments
    (b) small dipole moments
    (c) no dipole moment
    (d) large, permanent dipole moments

110. The relation between the compressibility and elastic constants of a cubic crystal is
    (a) $\beta = 3(C_{11} + 2C_{12})$
    (b) $\beta = 3(C_{11} + C_{12})$
    (c) $\beta = \dfrac{3}{2C_{11}}$
    (d) $\beta = \dfrac{3}{(C_{11} + 2C_{12})}$

111. The number of slip systems that exist in fcc metals are
    (a) 6
    (b) 8
    (c) 10
    (d) 12

112. The quantum of energy in an elastic wave is called as
    (a) phonon
    (b) photon
    (c) exciton
    (d) graviton

113. Superconductors are used for magnetic field shielding due to their strong
    (a) diamagnetism
    (b) paramagnetism
    (c) ferromagnetism
    (d) antiferromagnetism

114. The relation connecting the pressure and volume of an electron gas at 0 K is
    (a) $P = \dfrac{E_f(0)}{v}$
    (b) $P = -\dfrac{2}{5}\dfrac{E_f(0)}{v}$
    (c) $P = \dfrac{2}{5}\dfrac{E_f(0)}{v}$
    (d) $P = -\dfrac{E_f(0)}{v}$

115. In comparison with pure iron, iron–silicon alloy has
    (a) zero resistivity
    (b) very low resistivity
    (c) medium resistivity
    (d) high resistivity

116. The Fermi energy $E_f(0)$ is a function of
    (a) pressure
    (b) volume
    (c) temperature
    (d) susceptibility

117. The electrical conductivity of an intrinsic semiconductor is
    (a) $\sigma = (n\mu_e + p\mu_n)$
    (b) $(n\mu_e + p\mu_n)$
    (c) $\sigma = \dfrac{(n\mu_e + p\mu_n)}{|e|}$
    (d) $|e|(n\mu_e + p\mu_n)$

118. If a polymer sample contains an equal number of moles of species with degree of polymerization 1, 2, 3, 4, 5, 6, 7, 8 and 10 the weight with average degree of polymerization is
    (a) 4
    (b) 5
    (c) 6
    (d) 7

119. The Fermi energy E_f is a function of temperature and varies as

    (a) $E_f = E_f(0)\left[1 + \dfrac{\pi^2}{12}\left(\dfrac{K_T}{E_f(0)}\right)^2\right]$

    (b) $E_f = E_f(0)\left[1 - \dfrac{\pi^2}{12}\left(\dfrac{K_T}{E_f(0)}\right)^2\right]$

    (c) $E_f = E_f(0)\left[1 - \dfrac{\pi^2}{12}\left(\dfrac{K_T}{E_f(0)}\right)\right]$

    (d) $E_f = E_f(0)\left[1 - \dfrac{\pi^2}{12}\left(\dfrac{K_T}{E_f(0)}\right)\right]^2$

120. When the degree of polymerization is about 1000 mass per molecules, the substance becomes
    (a) a liquid
    (b) a true polymer
    (c) a paraffin
    (d) a greasy substance

121. In the mean field approximation, the susceptibility of ferromagnetic material above the Curie temperature has the form
(a) $\chi = \dfrac{C}{T+T_C}$
(b) $\chi = \dfrac{C}{T-T_C}$
(c) $\chi = C(T-T_C)$
(d) $\chi = \dfrac{C}{T_C}$

122. Those materials which consist of phases which are compounds of metals and non-metals are known as
(a) semiconductors
(b) superconductors
(c) ceramics
(d) conductors

123. An electron in Bloch state ψ_K moves through the crystal with a velocity $v = \dfrac{1}{\hbar}\nabla_K E(K)$. This velocity
(a) varies randomly
(b) decreases
(c) increases
(d) remains constant

124. The non-metallic materials which are used to withstand high temperature in different industrial processes and operations are known as
(a) glasses
(b) semiconductors
(c) refractories
(d) superconductors

125. The magnetic induction B inside a superconductor is always
(a) infinite
(b) positive
(c) zero
(d) negative

126. The critical current density J_c in superconductors is a function of
(a) H and T
(b) H only
(c) T only
(d) E and T

127. The energy of a spin wave is quantized and the unit of energy of a spin wave is called as
(a) phonon
(b) roton
(c) magnon
(d) photon

128. The critical magnetic field Hc required to destroy superconductivity is a function of
(a) temperature
(b) pressure
(c) volume
(d) electric field

129. A superconductor exhibits
    (a) zero conductivity
    (b) infinite resistivity
    (c) infinite conductivity
    (d) paramagnetism

130. When a metal undergoes the superconductivity transition, thermoelectricity
    (a) increases
    (b) decreases
    (c) remains same
    (d) vanishes

131. Infrared spectroscopy gives information concerning molecular
    (a) electronic frequencies
    (b) vibrational frequencies
    (c) rotational frequencies
    (d) structural data

132. When the metal undergoes the superconducting transition, the specific heat
    (a) vanishes
    (b) changes continuously
    (c) never changes
    (d) changes discontinuously

133. The number of peaks that appear in low resolution NMR spectrum of vinyl chloride is
    (a) 1
    (b) 6
    (c) 4
    (d) 3

134. The ratio of the volume of atoms to the total volume available in a simple cubic lattice is
    (a) 74%
    (b) 66%
    (c) 52%
    (d) 34%

135. A dielectric material has non-uniform polarization $\vec{p}$. The polarization volume charge density is given by
    (a) $|\vec{p}|^2$
    (b) $\dfrac{|\vec{p}|}{\varepsilon_0}$
    (c) $\vec{\nabla} \cdot \vec{p}$
    (d) $-\vec{\nabla} \cdot \vec{p}$

136. Madelung energy is the main contribution to the binding energy of
    (a) ionic crystals
    (b) covalent crystals
    (c) metals
    (d) inert gas solids

137. A Schottky defect in a crystal is an example of
    (a) a mixing atom
    (b) an extra atom
    (c) a colour centre
    (d) a dislocation

138. The packing fraction of a simple cubic crystal lattice is approximately
    (a) 0.74   (b) 0.68   (c) 0.52   (d) 0.34

139. The characteristic feature of the transition element is
    (a) a partly filled valence shell
    (b) an empty inner shell
    (c) an unfilled outer shell
    (d) a partly filled inner shell

140. The number of atoms per unit cell in the cubic diamond is
    (a) 4   (b) 6   (c) 7   (d) 8

141. The conductivity of a pure semiconductor is
    (a) proportional to temperature
    (b) rises exponentially with temperature
    (c) decreases exponentially with increasing temperature
    (d) independent of temperature

142. The slope of the graph, conductivity vs reciprocal temperature, in a semiconductor is
    (a) $-\dfrac{Eg}{2K}$ (b) $\dfrac{Eg}{2K}$ (c) $\dfrac{Eg}{K}$ (d) $\dfrac{KT}{Kg}$

143. A certain capacitor has a capacitance of 50 pF with air between its plates and 370 pF with a plastic between its plates. The dielectric constant of the plastic is
    (a) 74   (b) 7.4   (c) 0.012   (d) 0.12

144. An element can form a strongly magnetic solid only if its atom has
    (a) an incomplete valence shell
    (b) an incomplete inner shell
    (c) a vacant inner shell
    (d) a complete valence shell

145. Neel temperature is that temperature
    (a) above which the spins are not free
    (b) below which antiferromagnetic materials behave paramagnetic
    (c) above which susceptibility increases with increase of temperature
    (d) above which antiferromagnetic materials becomes paramagnetic

146. One of the following which gives Fermi energy at 0K in the increasing order is
    (a) Li, Cu, K, Na
    (b) Cu, Na, K, Li
    (c) K, Na, Li, Cu
    (d) Cu, Li, K, Na

147. The crunciers parameter (γ) is given by
    (a) $\gamma = \dfrac{\alpha V_0}{C_v \beta}$
    (b) $\gamma = \dfrac{\alpha \beta}{V_0 C_v}$
    (c) $\gamma = \dfrac{\alpha C_v}{V_0 \beta}$
    (d) $\gamma = \dfrac{C_v \beta}{\alpha V_0}$

148. The X-ray diffraction of a, b, c lattice does not contain the lines for which the sum of Miller indices h, k, l is
    (a) even   (b) odd   (c) zero   (d) real

149. The first Brillouin zone of a simple cubic lattice is
    (a) rhombic decahedron
    (b) truncated octahedron
    (c) parallelopiped
    (d) cube

150. If there are p atoms per unit cell the phonon dispersion relation will have
    (a) $3p$ acoustic phonon branches
    (b) $3p-3$ optical phonon branches and 3 acoustic phonon branches
    (c) $3p$ optical phonon branches
    (d) 6 acoustic branches

151. A metal insulator transition may occur when the nearest neighbour separation of the atoms a_0 becomes of the order of (a_0 is Bohr radius)
    (a) $2a_0$
    (b) a_0
    (c) $4a_0$
    (d) $\dfrac{a_0}{4}$

152. An exerton is a
    (a) bound electron-positron pair
    (b) electron bound to proton
    (c) photon
    (d) bound electron-hole pair

153. The paramagnetic susceptibility of alkali halides are neither temperature-dependent nor follow Curie–Weiss law. This was explained by Pauli using
    (a) Maxwell–Boltzmann statistics
    (b) Bose–Einstein statistics

(c) Fermi–Dirac statistics
(d) diamagnetic contribution

154. If $\cos\alpha$, $\cos\beta$ and $\cos\gamma$ are the direction cosines of any line in cartesian coordinates, then the following identity holds good
   (a) $\cos\alpha + \cos\beta + \cos\gamma = 1$
   (b) $\cos\alpha + \cos\beta - \cos\gamma = 1$
   (c) $\cos^2\alpha + \cos^2\beta + \cos^2\gamma = \sqrt{3}$
   (d) $\cos^2\alpha + \cos^2\beta + \cos^2\gamma = 1$

155. The optically active normal modes of a linear triatomic molecule of CO_2 are
   (a) 3   (b) 2   (c) 1   (d) 0

156. Out of the following materials the highest elastic material is
   (a) copper   (b) iron   (c) wood   (d) rubber

157. Alloy sheet steel has
   (a) high permeability and very small coercivity
   (b) high coercivity and small permeability
   (c) small coercivity and small permeability
   (d) high coercivity and high permeability

158. When air is replaced by a dielectric medium of constant k, the maximum force of attraction between two charges seperated by a distance d
   (a) decreases by k times
   (b) remains unchanged
   (c) increases by k times
   (d) becomes negative

159. The magnetic susceptibility χ_m
   (a) is independent of H for ferromagnetics
   (b) is independent of H for paramagnetics
   (c) is positive for all para, dia and ferromagnetics
   (d) is independent of temperature for paramagnetics

160. A material in which the ordered regions contain only a few atoms is termed
   (a) amorphous
   (b) glassy
   (c) polycrystalline
   (d) non-crystalline

161. A polycrystalline material always contains
    (a) crystals of different chemical composition
    (b) crystallite of the same composition but different structures
    (c) crystallites with different orientations
    (d) non-crystalline material

162. In case of diamond, the forbidden gap is about
    (a) 6.8 eV   (b) 1.2 eV   (c) 6 eV   (d) 10 eV

163. A beam of plasma-polarized electromagnetic radiation of frequency ω, electric field amplitude E_o, and polarization x is normally incident on a region of space containing a low density. Plasma ($P = 0$; n_o = electrons/volume). The conductivity as a function of frequency is

    (a) $\sigma = \dfrac{in_o e^2}{m\omega}$

    (b) $\sigma = \dfrac{n_o e^2}{m\omega}$

    (c) $\sigma = in_o e^2$

    (d) $\sigma = \dfrac{-e}{m} E'$

164. The dispersion relation for em waves in a plasma is given by $\omega^2(k) = \omega_p^2 + c^2 k^2$ where, ω_p the plasma frequency is defined as

    (a) $\omega_p^2 = \dfrac{Ne^2}{m}$

    (b) $\omega_p^2 = \dfrac{4\pi Ne^2}{m}$

    (c) $\omega_p^2 = \dfrac{4Ne^2}{m}$

    (d) $\omega_p^2 = \dfrac{\pi Ne^2}{m}$

    where symbols have their usual meaning.

165. Assume that the ionosphere consists of a uniform plasma for free electrons, neglecting collisions. The index of refraction for electromagnetic waves propagating in this medium in terms of the frequency is

    (a) $n = \sqrt{1 - \omega_p^2}$

    (b) $n = \sqrt{1 - \omega^2}$

    (c) $n = \sqrt{1 - \dfrac{\omega_p^2}{\omega^2}}$

    (d) $n = \sqrt{1 - \dfrac{\omega^2}{\omega_p^2}}$

    where, w_p is the plasma frequency.

166. The dispersion relation in the plasma is given by

(a) $\dfrac{\omega^2}{\omega_p^2}$ (b) $\dfrac{\omega^2}{C^2}$ (c) $\dfrac{\omega_p^2}{C^2}$ (d) $\dfrac{1}{C^2}(\omega^2 - \omega_p^2)$

167. The penetration depth of a very low frequency e-m wave into a plasma in which electrons are free to move is

(a) $\delta = \omega_p$ (b) $\delta = \dfrac{1}{\omega_p}$ (c) $\delta = \omega_p^2$ (d) $\delta \approx \left(\dfrac{1}{\omega_p}\right)^2$

168. The plasma bounded by vacuum, dielectric or another plasma can support
    (a) progressive waves   (b) resonance
    (c) surface waves   (d) shock waves
    which by definition are localized.

169. The linear MHD bulk waves (Alfien waves) are
    (a) non-compressive   (b) compressive
    (c) solenoid   (d) stationary

170. Ionic conductivity with temperature, composition and structure is the key to understand
    (a) optical properties
    (b) electrical and magnetic properties
    (c) transport properties
    (d) mechanical properties of crystals and glasses

171. Glasses are
    (a) isotropic materials   (b) anisotropic materials
    (c) solid solutions   (d) polymers

172. The formation of close-packed structure in polymers is prevented by
    (a) metallic bond chains   (b) covalent bond chains
    (c) ionic bond chains   (d) hydrogen bond chains

173. The penetration depth of a very low frequency e-m waves into a plasma in which electrons are free to move for $n_o = 10^{14}$ cm^{-3} is
    (a) 5.3 cm   (b) 0.53 cm
    (c) 0.053 cm   (d) 0.0053 cm

174. The conductivity of glass due to alkali cations is
  (a) $\sigma = nze\mu$
  (b) $\sigma = ze\mu$
  (c) $\sigma = -nze\mu$
  (d) $\sigma = n\mu$

175. Glasses possess
  (a) equilibrium ordered phases
  (b) equilibrium disordered phases
  (c) non-equilibrium ordered phases
  (d) non-equilibrium disordered phases

176. The main advantage of a polymer electrolyte stem from its remarkable
  (a) transport properties
  (b) mechanical properties
  (c) optical properties
  (d) electrical properties

177. MHD surface waves can be
  (a) compressive
  (b) non-compressive
  (c) compressive or non-compressive
  (d) stationary

178. A cold collisionless plasma is characterized by the dielectric permittivity $\varepsilon_p = 1 - \dfrac{\omega_p^2}{\omega^2}$ which is
  (a) zero for low frequencies
  (b) negative for low frequencies
  (c) positive for low frequencies
  (d) imaginary for low frequencies

179. In a cold plasma, for pure surface waves (symmetric or non-symmetric) only
  (a) one mode
  (b) two modes
  (c) three modes
  (d) four modes is possible

180. The transverse t waves arise in cyclotron resonance due to
  (a) exponential distribution function
  (b) parabolic distribution function
  (c) asymmetry of the distribution function
  (d) symmetry of the distribution function

181. Small amplitude MHD waves in uniform plasmas are known to be
 (a) non-dispersive (b) dispersive
 (c) linear (d) surface waves

182. The electrical conductivity of AgBr steeply increases to the extraordinary value of
 (a) 0.5/ohm/cm (b) 1.0/ohm/cm
 (c) 1.5/ohm/cm (d) 2.0/ohm/cm

183. A plasma is essentially a gas consisting of
 (a) neutral atoms (b) charged particles, electrons and ions
 (c) molecules (d) radicals

184. In neutral gas the forces are
 (a) very weak (b) long range
 (c) van der Waals forces (d) very strong but of short range

185. The Debye length, a parameter which is of fundamental importance in a plasma is given by
 (a) $\lambda^2 = \dfrac{\varepsilon_0 k}{(n_0 e^2)}$
 (b) $\lambda^2 = \varepsilon_0 kT$
 (c) $\lambda = \left[\dfrac{\varepsilon_0 kT}{n_0}\right]^{\frac{1}{2}}$
 (d) $\lambda = \left[\dfrac{\varepsilon_0 kT}{n_0 e^2}\right]^{\frac{1}{2}}$

186. In a plasma, Kruskal–Shafranor's stability condition is
 (a) $\left|\dfrac{B_\theta}{B_z}\right| < \dfrac{2\pi a}{L}$
 (b) $\left|\dfrac{B_\theta}{B_z}\right| = \dfrac{2\pi a}{L}$
 (c) $\left|\dfrac{B_\theta}{B_z}\right| > \dfrac{2\pi a}{L}$
 (d) $\left|\dfrac{B_\theta}{B_z}\right| = 0$

187. The limit where the plasma pressure becomes negligible, the slow wave disappears while the fast wave propagates in a spherically symmetric fashion is known as
 (a) acoustic wave
 (b) non-compresssional wave
 (c) compressional Aljven's wave
 (d) magnetosonic waves

188. Typical frequencies for electron cyclotron heating are of the order of
    (a) 20–100 Hz
    (b) 20–100 KHz
    (c) 20–100 MHz
    (d) 20–100 GHz

189. In fast-ion conductors, the high electrical conductivity is comparable with
    (a) solid electrolytes
    (b) liquid electrolytes
    (c) superconductors
    (d) composite materials

190. As the ionic radius becomes higher, the conductivity becomes
    (a) zero
    (b) infinity
    (c) lower
    (d) higher

191. The surface structure can be studied by
    (a) X-ray topography techniques
    (b) spectroscopic techniques
    (c) diffraction techniques
    (d) chemical techniques

192. The electrical conductivity in supersonic solids is given by
    (a) $\sigma = G\left(\dfrac{l}{A}\right)$ ohm^{-1}cm^{-1}
    (b) $\sigma = \left(\dfrac{l}{A}\right)$ ohm^{-1}cm^{-1}
    (c) $\sigma = G\left(\dfrac{A}{l}\right)$ ohm^{-1}cm
    (d) $\sigma = \left(\dfrac{A}{l}\right)$ ohm^{-1}cm

193. In general, plasma is
    (a) electrically neutral overall
    (b) positively charged particles
    (c) negatively charged particles
    (d) neutral particles

194. The existence of charged particles in plasma means
    (a) it can be treated simply as an ordinary gas
    (b) it cannot react to electric current
    (c) it cannot support on electric current and react to electric and magnetic fields
    (d) it cannot react to electric and magnetic fields

195. In plasma, the forces are
    (a) van der Waals forces   (b) Coulomb forces
    (c) strong and short range   (d) zero

196. Sausage instability (plasma deformation)
    (a) decreases magnetic pressure
    (b) decreases the perturbation
    (c) creates no change
    (d) increases magnetic pressure and hence increases perturbation

197. In real life, plasmas are
    (a) not usually the spatially homogeneous systems
    (b) usually the spatially homogeneous systems
    (c) not usually the temporally homogeneous systems
    (d) usually the temporally homogeneous systems

198. The parameters of a tokamak are generally such that the electron plasma frequencies and cyclotron frequencies are of the same order of magnitude
    (a) outside the plasma
    (b) far away from plasma
    (c) near the corner of the plasma
    (d) near the centre of the plasma

199. From the crystallographic point of view, a perfect crystal of an ionic compound would be
    (a) a good conductor   (b) a semiconductor
    (c) an insulator   (d) a superconductor

200. Lithium is a very promising material for high energy density batteries because of its
    (a) heavy weight   (b) high electrochemical potential
    (c) alkali nature   (d) phase transition

201. To obtain partially covalent bonding in compounds, it is necessary to have
    (a) highly polarizable cations
    (b) highly polarizable anions
    (c) both highly polarizable cations and anions
    (d) chargeless particles

202. The macroscopic property of superionic solids can be studied through
    (a) ion dynamics
    (b) ion transport
    (c) thermodynamic properties
    (d) structural characterization

203. On the basis of 2-fold rotational symmetry, two-dimensional lattice can be classified as
    (a) square
    (b) rectangular
    (c) hexagonal
    (d) oblique

204. The Bragg's law is represented as
    (a) $2K \cdot G + G = 0$
    (b) $2K \cdot G + G^2 = 0$
    (c) $2K \cdot G - G = 0$
    (d) $2K \cdot G - G^2 = 0$

205. The melting temperature for the ionic crystal NaCl is
    (a) 601°C
    (b) 701°C
    (c) 801°C
    (d) 901°C

206. If the potential energy function is expressed as $u(r) = \dfrac{-\alpha}{r^6} + \dfrac{\beta}{r^{12}}$, then the minimum potential energy is
    (a) $\dfrac{-\alpha^2}{4\beta}$
    (b) $\dfrac{\alpha^2}{4\beta}$
    (c) $\dfrac{-\alpha}{4\beta}$
    (d) $\dfrac{\alpha}{4\beta}$

207. For a cubic lattice, $C_{11} - C_{12}$ is equal to
    (a) $2C_{22}$
    (b) $2C_{44}$
    (c) C_{22}
    (d) C_{44}

    where symbols have their usual meaning.

208. The thermal conductivity of metals is given by $K = LT\sigma$ where L is a constant known as Lorentz number and its value is
    (a) $\dfrac{1}{3}\left(\dfrac{K}{e}\right)^2$
    (b) $\dfrac{\pi}{3}\left(\dfrac{K}{e}\right)$
    (c) $\pi\left(\dfrac{K}{e}\right)$
    (d) $\dfrac{\pi^2}{3}\left(\dfrac{K}{e}\right)^2$

209. An electron in Block state ψ_k moves through the crystal with a velocity
    (a) $v = \dfrac{1}{\hbar}\nabla_k E(k)$
    (b) $v = \nabla_k E(k)$
    (c) $v = \hbar \nabla_k E(k)$
    (d) $v = \dfrac{1}{\hbar}\nabla_k^2 E(k)$

210. In the mean field approximation, the susceptibility of ferromagnet above Curie temperature has the form

(a) $\chi = \dfrac{C}{T+T_C}$
(b) $\chi = \dfrac{C}{T-T_C}$
(c) $\chi = \dfrac{1}{T+T_C}$
(d) $\chi = \dfrac{1}{T-T_C}$

211. The relation between the crystal axis in oblique lattice (oblique system, point group 1, 2) is

(a) $a = b, \phi = 90$
(b) $a \neq b, \phi = 90$
(c) $a = b, \phi = 120$
(d) $a \neq b, \phi \neq 120$

212. The density or the packing fraction in the case of simple cubic structure is about

(a) $\dfrac{\pi}{2}$
(b) $\dfrac{\pi}{4}$
(c) $\dfrac{\pi}{6}$
(d) $\dfrac{\pi}{8}$

213. Covalent crystals have
(a) zero electrical resistance
(b) very low electrical resistance
(c) constant electrical resistance
(d) variable electrical resistance

214. The binding of molecular hydrogen is a simple example of
(a) covalent bond
(b) ionic bond
(c) metallic bond
(d) dispersive bond

215. The relation between the compressibility and elastic constants of a cubic crystal is

(a) $\beta = \dfrac{1}{C_{11}+2C_{12}}$
(b) $\beta = \dfrac{2}{C_{11}+2C_{12}}$
(c) $\beta = \dfrac{3}{C_{11}+2C_{12}}$
(d) $\beta = \dfrac{4}{C_{11}+2C_{12}}$

216. The energy spectrum of one dimension free electron gas is discrete with energy level separation depending on

(a) n
(b) L
(c) $\dfrac{n^2}{L^2}$
(d) $\dfrac{n}{L}$

217. In the presence of an electric field, an electron moves in K-space according to the relation

(a) $K = \dfrac{eE_0}{\hbar}$
(b) $K = \dfrac{-eE_0}{\hbar}$
(c) $K = \dfrac{E_0}{\hbar}$
(d) $K = \dfrac{-E_0}{\hbar}$

218. The diamagnetic susceptibility (χ) of N atoms of atomic number is

(a) $\chi = \dfrac{-NZe^2}{6mc^2}<r^2>$
(b) $\chi = \dfrac{NZe^2}{6mc^2}<r^2>$
(c) $\chi = \dfrac{-N}{6mc^2}<r^2>$
(d) $\chi = \dfrac{N}{6mc^2}<r^2>$

where, $<r^2>$ is the mean square atomic radius.

219. A superconductor is a perfect diamagnet, with the magnetic induction $B = 0$. This is
(a) DC Josephson effect
(b) superconductor tunnelling
(c) Meissner effect
(d) AC Josephson effect

220. The critical magnetic field for aluminium is 7.9×10^3 amp/metre the critical current which can flow through a thin long aluminium superconducting wire of diameter 10^{-3} metre is
(a) 20.806 amp
(b) 21.806 amp
(c) 24.806 amp
(d) 23.806 amp

221. Magnetic flux through a superconducting ring is quantized and the effective unit charge is
(a) e
(b) $2e$
(c) $3e$
(d) $4e$

222. The space lattice of caesium chloride structure is
(a) hcp
(b) bcc
(c) fcc
(d) scc

223. Superionic conductors are generally formed with
(a) electrovalent bonding
(b) covalent bonding
(c) homopolar bonding
(d) metallic bonding

224. Magnon is a quantized
 (a) spinwave
 (b) thermal wave
 (c) photon
 (d) neutron

225. The curve molar susceptibility vs temperature for MnF_2 shows a peak at 720 K. This temperature is called
 (a) Curie temperature
 (b) Weiss temperature
 (c) Langevin temperature
 (d) Neel temperature

ANSWERS

1. (d)	2. (b)	3. (b)	4. (b)	5. (b)
6. (d)	7. (b)	8. (b)	9. (d)	10. (a)
11. (b)	12. (d)	13. (c)	14. (d)	15. (d)
16. (c)	17. (d)	18. (a)	19. (b)	20. (a)
21. (a)	22. (b)	23. (c)	24. (b)	25. (a)
26. (c)	27. (d)	28. (a)	29. (a)	30. (c)
31. (b)	32. (c)	33. (d)	34. (a)	35. (c)
36. (a)	37. (b)	38. (b)	39. (a)	40. (c)
41. (a)	42. (a)	43. (c)	44. (a)	45. (a)
46. (a)	47. (c)	48. (c)	49. (a)	50. (d)
51. (d)	52. (b)	53. (c)	54. (b)	55. (d)
56. (a)	57. (a)	58. (a)	59. (d)	60. (c)
61. (c)	62. (b)	63. (c)	64. (c)	65. (b)
66. (c)	67. (b)	68. (b)	69. (c)	70. (a)
71. (b)	72. (b)	73. (b)	74. (c)	75. (a)
76. (b)	77. (d)	78. (a)	79. (d)	80. (b)
81. (a)	82. (a)	83. (c)	84. (d)	85. (b)
86. (b)	87. (d)	88. (c)	89. (b)	90. (d)
91. (c)	92. (d)	93. (b)	94. (b)	95. (a)
96. (b)	97. (a)	98. (a)	99. (b)	100. (c)
101. (b)	102. (d)	103. (b)	104. (c)	105. (a)
106. (a)	107. (c)	108. (c)	109. (d)	110. (d)
111. (d)	112. (a)	113. (a)	114. (c)	115. (d)
116. (c)	117. (a)	118. (d)	119. (b)	120. (b)

121. (b)	122. (c)	123. (d)	124. (c)	125. (c)
126. (a)	127. (c)	128. (a)	129. (c)	130. (d)
131. (b)	132. (d)	133. (d)	134. (c)	135. (d)
136. (a)	137. (a)	138. (c)	139. (a)	140. (a)
141. (a)	142. (a)	143. (b)	144. (b)	145. (d)
146. (c)	147. (b)	148. (d)	149. (d)	150. (b)
151. (c)	152. (d)	153. (d)	154. (d)	155. (a)
156. (b)	157. (c)	158. (a)	159. (d)	160. (c)
161. (c)	162. (c)	163. (a)	164. (b)	165. (c)
166. (d)	167. (b)	168. (c)	169. (a)	170. (c)
171. (a)	172. (b)	173. (c)	174. (a)	175. (d)
176. (b)	177. (c)	178. (b)	179. (a)	180. (c)
181. (a)	182. (c)	183. (b)	184. (d)	185. (d)
186. (a)	187. (c)	188. (d)	189. (b)	190. (c)
191. (a)	192. (a)	193. (a)	194. (c)	195. (b)
196. (d)	197. (a)	198. (d)	199. (c)	200. (b)
201. (c)	202. (b)	203. (d)	204. (b)	205. (c)
206. (a)	207. (b)	208. (d)	209. (a)	210. (b)
211. (b)	212. (c)	213. (d)	214. (a)	215. (c)
216. (c)	217. (b)	218. (a)	219. (c)	220. (c)
221. (b)	222. (d)	223. (a)	224. (a)	225. (d)

5

QUANTUM MECHANICS

FORMULAE

1. **Rayleigh Jeans law**

$$u_\upsilon = \frac{8\pi}{c^3}\upsilon^2 kT$$

Here,

υ = frequency of the oscillator
kT = average energy of the oscillator
u_υ = energy density

2. **Planck's formula**

$$u_\upsilon = \frac{8\pi h \upsilon^3}{c^3} \frac{1}{\exp\left(\frac{h\upsilon}{kT}-1\right)}$$

Photon Energy

$$E = h\upsilon$$

i.e., Energy = (Planck's constant) × frequency

de Broglie wavelength

$$\lambda = \frac{h}{p}$$

Here, p is the momentum of the particle.

3. Uncertainty Principle

$$\Delta x \, \Delta P_x \geq h$$

Here,

Δx = uncertainty in position
ΔP_x = uncertainty in momentum

4. Operator Representation Energy

$$E = i\hbar \frac{\partial}{\partial t}$$

Momentum

$$P = -i\hbar \frac{\partial}{\partial x}$$

5. Schrodinger equation

$$\frac{-\hbar^2}{2m} \nabla^2 \psi + V\psi = i\hbar \frac{\partial \psi}{\partial t}$$

Here,

ψ = wave function
m = mass of the particle

6. Commutator Relation

$$[xP_x - P_x x] = [x, P_x] = i\hbar$$

$$[x, y] = [x, z] = 0$$

$$[x, P_y] = [y, P_z] = 0$$

$$[P_x, P_y] = [P_z, P_x] = 0$$

7. Equation of Continuity

$$\frac{\partial \rho}{\partial t} + \nabla \cdot J = 0$$

Here,

ρ = probability density
J = probability current density

$$\rho = \psi^*\psi$$

$$J = \frac{i\hbar}{2m}\left[\psi^*\nabla\psi^* - \psi^*\nabla\psi\right]$$

8. Expectation Values

$$\langle x \rangle = \frac{\iiint x\psi^*\psi\, d\tau}{\iiint \psi^*\psi\, d\tau}$$

9. One dimensional time-independent Schrodinger equation

$$\frac{d^2\psi}{dx^2} + \frac{2m}{\hbar^2}[E - V(x)]\psi(x) = 0$$

Here,

E = energy
$V(x)$ = potential
m = mass of the particle

10. Particle in a box is represented by

$$\text{Energy} = \frac{n^2\hbar^2\pi^2}{2\mu a^2}$$

Here, μ is the mass of the particle.

11. Wave function

$$\psi_n = \sqrt{\frac{2}{a}}\sin\left(\frac{n\pi x}{a}\right), \quad 0 < x < a$$

Here, a is the length of the box.

12. Linear harmonic oscillator

$$\text{Energy} = \left(n + \frac{1}{2}\right)\hbar\omega$$

Wave function

$$\psi_n(x) = N_n \exp\left(-\frac{1}{2}\xi^2\right) H_n(\xi)$$

$$N_n = \left[\frac{\alpha}{\sqrt{\pi}\, 2^n\, n!}\right]^{\frac{1}{2}}$$

$$\xi = \alpha x = \left(\frac{m\omega}{\hbar}\right)^{\frac{1}{2}} x$$

where, H_n = Hermite polynomial.

13. Barrier Transmission problem

$$R = \frac{J_{\text{ref}}}{J_{\text{inc}}}$$

$$= \frac{|B|^2}{|A|^2}$$

$$= \frac{(K - K_1)^2}{(K_3 + K_2)^2}$$

$$T = \frac{J_{tr}}{J_{\text{inc}}}$$

$$= \frac{K_1 |C|^2}{K |A|^2}$$

$$= \frac{4 K K_1}{(K + K_1)^2}$$

Here,

R = reflection coefficient
T = transmission coefficient

$$K_1 = \left[\frac{2m}{\hbar^2}(V_o - E)\right]^{\frac{1}{2}}$$

Here,

m = mass of the particle
V_o = the potential of the barrier
E = the energy of the particle

14. The Hydrogen atom

The probability of finding electron between r and $r + dr$ is

$$P(r)dr = |\psi_{100}|^2 4\pi r^2 dr$$

$$= \left(\frac{4Z}{a_o}\right)^3 r^2 \exp\left(\frac{-2Zr}{a_o}\right) dr$$

Here,

a_o = Bohr radius
  = 0.53 Å
Z = atomic number

15. Angular Momentum I

Angular momentum of the particle is

$$L = r \times p$$

Here,

r = position of the particle
p = linear momentum of the particle

$$L_x = yP_z - yP_y$$

$$L_x = -i\hbar\left(y\frac{\partial}{\partial z} - z\frac{\partial}{\partial y}\right)$$

$$L_y = zP_x - xP_z = -i\hbar\left(z\frac{\partial}{\partial x} - x\frac{\partial}{\partial z}\right)$$

$$L_z = xP_y - yP_x = -i\hbar\left(x\frac{\partial}{\partial y} - y\frac{\partial}{\partial x}\right)$$

If it is **Spherical Polar Coordinates** then

$$L_x = -i\hbar\left(\sin\phi\frac{\partial}{\partial\theta} + \cot\theta\cos\phi\frac{\partial}{\partial\phi}\right)$$

$$L_y = -i\hbar\left(-\cos\phi\frac{\partial}{\partial\theta} + \cot\theta\sin\phi\frac{\partial}{\partial\phi}\right)$$

$$L_z = -i\hbar\frac{\partial}{\partial\phi}$$

Eigen values of L^2

$$L^2 Y(\theta,\phi) = \lambda\hbar^2 Y(\theta,\phi)$$

$$L^2 = \lambda\hbar^2$$

$$= l(l+1)\hbar^2$$

16. Commutation Relations

$$[L_x, L_y] = i\hbar L_z$$

$$[L_y, L_z] = i\hbar L_x$$

$$[L_z, L_y] = i\hbar L_y$$

$$[L^2, L_x] = [L^2, L_y] = [L^2, L_z] = 0$$

17. Ladder operators

$$L_+ = L_x + iL_y$$

$$L_- = L_x - iL_y$$

$$L_+ = [L_z, L_+]$$

$$-L_- = [L_z, L_-]$$

18. Paulis Spin Matrices

$$\delta_x = \begin{pmatrix} 0 & 1 \\ 1 & 0 \end{pmatrix}$$

$$\delta_y = \begin{pmatrix} 0 & -i \\ i & 0 \end{pmatrix}$$

$$\delta_z = \begin{pmatrix} 1 & 0 \\ 0 & -1 \end{pmatrix}$$

19. Clebsch–Gordon Coefficients

for $J_2 = \frac{1}{2}$; $\left\langle J_1 = \frac{1}{2} m_1 \, m_2 \middle| J_1 = \frac{1}{2}, J_m \right\rangle$

J	$M_2 = \frac{1}{2}$	$M_2 = -\frac{1}{2}$
$J_1 + \frac{1}{2}$	$\sqrt{\dfrac{J_1 + m + \frac{1}{2}}{2J_1 + 1}}$	$\sqrt{\dfrac{J_1 - m + \frac{1}{2}}{2J_1 + 1}}$
$J_1 - \frac{1}{2}$	$-\sqrt{\dfrac{J_1 - m + \frac{1}{2}}{2J_1 + 1}}$	$\sqrt{\dfrac{J_1 + m + \frac{1}{2}}{2J_1 + 1}}$

$$J_2 = 1 \langle J_1 | m_1, m_2 | J_1 1 J_m \rangle$$

J	$m_2 = 1$	$m_2 = 0$	$m_2 = -1$
$J_1 + 1$	$\sqrt{\dfrac{(J_1+m)(J_1-m+1)}{(2J_1+1)(2J+2)}}$	$\sqrt{\dfrac{(J_1-m+1)(J_1+m+1)}{(2J_1+1)(J_1+1)}}$	$\sqrt{\dfrac{(J_1-m)(J_1-m+1)}{(2J_1+1)(2J_1+2)}}$
J_1	$-\sqrt{\dfrac{(J_1+m)(J_1-m+1)}{2J_1(J_1+1)}}$	$\dfrac{m}{\sqrt{J_1(J_1+1)}}$	$\sqrt{\dfrac{(J_1-m)(J_1+m+1)}{2J_1(J_1+1)}}$
$J_1 - 1$	$\sqrt{\dfrac{(J_1-m)(J_1-m+1)}{2J_1(2J_1+1)}}$	$\sqrt{\dfrac{(J_1+m+1)(J_1+m)}{2J_1(2J_1+1)}}$	$\sqrt{\dfrac{(J_1+m+1)(J_1+m)}{2J_1(2J_1+1)}}$

20. Scattering cross section

$$\sigma(\theta,\phi)d\Omega = \frac{\text{Number of particles scattered into solid angle } d\Omega \text{ per unit time}}{\text{Incident intensity } I_o}$$

Here,

$d\Omega$ = solid angle
θ = the angle of scattering
ϕ = the azimuthal angle

The wave function of the outgoing scattered wave is

$$\psi_{out} = f(\theta)\frac{e^{ikr}}{r}$$

Here, $f(\theta)$ = scattering amplitude.

Scattering cross section is

$$d\sigma = d\Omega \, |f(\theta,\phi)|^2$$

Here,

$d\Omega$ = solid angle
$|f(\theta,\phi)|$ = scattering amplitude

Scattering cross section is

$$\sigma = 4\pi a^2$$

21. WKB approximation

Potential-slowly varying potential

$$\left|\frac{1}{K(x)}\frac{dK}{Kx}\right| << K,$$

where, $K(x) = \frac{1}{\hbar}P(x) = \left(\frac{2m}{\hbar^2}(E - V(x))\right)^{\frac{1}{2}}$.

In terms of wave number

$$K(x) = \frac{2\pi}{\lambda(x)}$$

WKB solution for
$$\frac{d^2\psi}{dx^2} + 2m[E - V(x)]\psi(x) = 0$$

is
$$\psi(x) = \frac{D_1}{\sqrt{K}} \int_e^x K(x)dx + \frac{D_2}{\sqrt{K}} - \int_e^x K(x)dx$$

Connection Formulae

$\alpha > 0$

$$\frac{2}{\sqrt{K}} \sin\left[\int_x^a K(x)dx + \frac{\pi}{4}\right] <-> \frac{1}{\sqrt{K}} \exp\left[-\int_a^x K(x)dx\right]$$

$$\frac{1}{\sqrt{K}} \cos\left[\int_x^a K(x)dx + \frac{\pi}{4}\right] <-\gg \frac{1}{\sqrt{K}} \exp\left[\int_a^x K(x)dx\right]$$

$$\frac{1}{\sqrt{K}} \exp\left[-\int_x^a K(x)dx\right] <-\gg \frac{2}{\sqrt{K}} \sin\left[\int_a^x K(x)dx + \frac{\pi}{4}\right]$$

$$\frac{1}{\sqrt{K}} \exp\left[\int_x^a K(x)dx\right] <-> \frac{1}{\sqrt{K}} \cos\left[\int_a^x K(x)dx + \frac{\pi}{4}\right]$$

22. WKB Solution for Bound State Energy Level for Potential

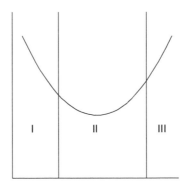

$$\psi_1 = \frac{1}{\sqrt{K}} \exp\left[-\int_x^{x_1} K(x)\, dx\right]$$

$$\psi_{11} = \frac{2}{\sqrt{K}} \cos\theta \cos\left(\int_x^{x_2} K\,dx + \frac{\pi}{4}\right) + \frac{2}{\sqrt{K}} \sin\theta \sin\left(\int_x^{x_2} K\,dx + \frac{\pi}{4}\right)$$

23. WKB Approximation to Barrier Penetration Problem

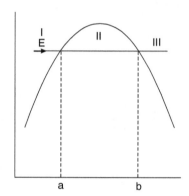

$$\psi_1 = \frac{A}{\sqrt{K}} e^{-i\int_x^a K(x)\,dx} + \frac{B}{\sqrt{K}} e^{i\int_x^a K(x)\,dx}$$

In region II

$$\psi_{II} = \frac{F}{\sqrt{K}} e^{i\frac{\pi}{4}} e^{\int_x^b K\,dx}$$

In region III

$$\psi_{III} = \frac{F}{\sqrt{K(x)}} \exp\left(i\int_b^x K(x)\,dx\right)$$

24. Time-independent perturbation theory

First order correction is

$$a_K^{(1)} = \frac{H'_{Kn}}{(E_n - E_K)}$$

Here, E_n and E_K are the energy states of n and E.

Second order correction is

$$a_K^{(2)} = \sum \frac{H'_{Km} H'_{mn}}{(E_n - E_K)(E_n - E_m)} - \frac{H'_{Kn} H'_{nn}}{(E_n - E_K)^2}$$

25. Time-dependent perturbation theory

Probability amplitude is given by

$$C_f(t) = \frac{1}{i\hbar} V_{fi} \int_0^t \exp(i\omega_{fi} t) dt$$

$$= -V_{fi} \left[\frac{\exp(i\omega_{fi} t) - 1}{\hbar \omega_{fi}} \right]$$

Probability is given by

$$P_{f(t)} = |C_f(t)|^2 = |V_{fi}|^2 \frac{1}{\hbar^2} \frac{\sin^2\left(\frac{\omega_{fi} t/2}{2}\right)}{\left(\frac{\omega_{fi}}{2}\right)^2}$$

where, $\omega_{fi} = E_f - \frac{E_i}{\hbar}$

26. Fermi's Golden Rule

Transition probability per unit time is

$$T = \frac{d\omega}{dt} = \frac{2\pi}{\hbar} |V_{fi}|^2 \rho_f(E_i)$$

Here,

$\rho_f(E_i)$ = number of energy states
$|V_{fi}|$ = potential

Tunnel Effect

$$\text{Transmission coefficient} = \frac{16 e^{-2\beta a}}{\left[1 + \left(\frac{\beta}{\alpha}\right)^2\right]\left[1 + \left(\frac{\alpha}{\beta}\right)^2\right]}$$

$$\alpha = 2mE/h^2; \quad \beta = \frac{2m(E-V_0)}{h^2}$$

Here,

α = length of the barrier
m = mass of the particle
E = energy of the particle
V_0 = potential of the barrier

Energy of the harmonic oscillator is given by

$$E_n = \left(n + \frac{1}{2}\right)h\upsilon$$

Here,

n = vibrational quantum number (0, 1, 2, ...)
υ = frequency of the oscillator
h = Planck's constant

Ehrenfest's Principle

$$[H,P] = i\hbar \frac{\partial V}{\partial x}$$

Here,

H = Hamiltonian of the particle
P = momentum of the particle
V = potential energy of the particle

Differential scattering cross section

$$\sigma(\theta) = |f(\theta)|^2$$

Identical Particles

The wave function for bosons is

$$\psi(1, 2, 3,..., S, N) = \psi(1, 2, 3,..., N)$$

The wave function for fermions is

$$\psi(1, 2,..., S, N) = -\psi(1, 2,..., N)$$

Pauli Spin Operators are as follows.

$$\hat{S}_x\hat{S}_y - \hat{S}_y\hat{S}_x = i\hbar\hat{S}_z$$
$$\hat{S}_z\hat{S}_x - \hat{S}_x\hat{S}_z = i\hbar\hat{S}_y$$
$$\hat{S}_y\hat{S}_z - \hat{S}_z\hat{S}_y = i\hbar\hat{S}_x$$

where, $\hat{S}_x, \hat{S}_y, \hat{S}_z$ are the operators,

and $S_x = \begin{pmatrix} 0 & \frac{\hbar}{2} \\ \frac{\hbar}{2} & 0 \end{pmatrix}; S_y = \begin{pmatrix} 0 & \frac{i\hbar}{2} \\ \frac{i\hbar}{2} & 0 \end{pmatrix}; S_z = \begin{pmatrix} \frac{\hbar}{2} & 0 \\ 0 & -\frac{\hbar}{2} \end{pmatrix}$

27. **Klien–Gordon Equation**

$$C^2\left[-i\hbar\nabla + \frac{e}{c}A\right]\left[-i\hbar\nabla + \frac{e}{c}A\right]\psi(r,t)$$
$$+m_0^2c^4\psi(r,t) = \left(i\hbar\frac{\partial}{\partial t} + e\phi\right)\left(i\hbar\frac{\partial}{\partial t} + e\phi\right)\psi(r,t)$$

Here,
- c = velocity of light
- e = charge of the particle
- A, ϕ = electromagnetic potentials

MULTIPLE CHOICE QUESTIONS

1. For a particle in a one-dimensional square well, the energy difference between successive levels is proportional to
   (a) n^2 (b) n (c) $n^2 - n$ (d) $\frac{1}{n^2}$

2. The ratio of quantum mechanical and classical cross section for scattering by a rigid sphere at high energy is
   (a) 4   (b) 3   (c) 2   (d) 1

3. Which quantum numbers determine the Stark effect in hydrogen?
   (a) n, l, s, j (b) n, l, s (c) n, l (d) n only

4. The wave function $u_n(x)$ for a linear harmonic oscillator has the property $\int_{-\infty}^{\infty} u_n(x) \times u_m(x) = 0$ where, m is equal to
   (a) $(n+1)$
   (b) $(n-1)$
   (c) values other than (a) and (b)
   (d) zero

5. The uncertainty product $\Delta x \cdot \Delta p$ for a harmonic oscillator is
   (a) $\dfrac{h}{2\pi}$
   (b) $\dfrac{h}{4\pi}$
   (c) $\dfrac{nh}{2\pi}$
   (d) $\left(n+\dfrac{1}{2}\right)\dfrac{h}{2\pi}$

6. If ϕ, ψ_i and ρ are the electronic, nuclear vibrational and rotational wave functions according to the Born–Oppenheimer approximation, the total wave function ψ of the molecule is given by
   (a) $\psi = \phi + \psi_i + \rho$
   (b) $\psi = \phi + (\psi_i \rho)$
   (c) $\psi = \phi \psi_i \rho$
   (d) $\psi = (\phi \psi_i) + \rho$

7. In non-relativistic quantum mechanics, the plane velocity of matter waves associated with a particle of velocity v is
   (a) $\dfrac{v}{2}$
   (b) v
   (c) $2v$
   (d) unrelated to v

8. The x-coordinate of a particle is measured with rms uncertainty $\Delta x = a$. From this alone, it follows that the rms uncertainty in P_x, viz., ΔP_x is
   (a) zero
   (b) determined uniquely in terms of $\hbar$ and a
   (c) $\dfrac{\hbar}{4a}$
   (d) anywhere between $\dfrac{\hbar}{2a}$ and infinity

9. A hydrogen atom is in 1 P state. Noting that both the proton and electron carry spin quantum number $\dfrac{1}{2}$, the net angular momentum quantum number J of the entire hydrogen atom is
   (a) can take any of the values 0, 1, 2
   (b) can never be zero
   (c) is necessarily one
   (d) is necessarily two

10. The ground state wave function ψ of hydrogen atom in spherical polar coordinates has no angular dependence but only radial dependence. Then ψ is an eigen function of
    (a) L_x, L_y, L_z simultaneously
    (b) L_z but not of L_x and L_y
    (c) none of the L_x, L_y and L_z
    (d) P_x

11. For the two-state physical system, an observable A is represented by the matrix $A = \begin{bmatrix} 1 & -1 \\ -1 & 1 \end{bmatrix}$, then
    (a) an exact measurement of A can yield any number between $-\infty$ to $+\infty$ is the possible result
    (b) $0 \leq \langle A \rangle \leq 2$ for every physical state of the system
    (c) $\langle A \rangle = 1$ for every physical state of the system
    (d) $0 \leq \langle A \rangle \leq 1$ for every physical state of the system

12. Let E'_n ($n = 0, 1, 2,...$) be the eigen values for a particle of mass m placed in an anharmonic potential $V(x) = \frac{1}{2} m w^2 x^2 + ax^4 (a > 0)$.
    Let $E_n = \left(n + \frac{1}{2}\right) \hbar$. Then according to the 1st order permutation theory
    (a) $E'_0 = E_0$
    (b) $E'_0 > E_0$
    (c) $E'_0 < E_0$
    (d) $E'_n < E_n$ for all n

13. The effect of an operator $f(x)$ as an arbitrary state $\psi_a = |a\rangle$ with components $\langle x'|a\rangle = \psi_a(x')$ is
    (a) $\int dx' |x> f(x') \psi_a(x')$
    (b) $\int dx' f(x') \psi_a(x')$
    (c) $\int dx' |x'> f(x')$
    (d) $i\hbar \int dx' \frac{df(x')}{dx'} |x'\rangle$

14. Pick out the set of vectors which are linearly independent.
    (a) (1, 0, – 1), (0, 1, 0), (1, 0, 1)
    (b) (1, 0, 1), (0, 1, 1), (1, 1, 2)
    (c) (1, 0, – 1), (0, 1, 0), (– 1, 1, 1)
    (d) (1, 1, 1), (– 1, 0, 0), (0, 1, 1)

15. The Schrodinger picture is characterized by
    (a) moving-state vectors and fixed operators
    (b) moving-state vectors and moving operators
    (c) fixed-state vectors and fixed operators
    (d) fixed-state vectors and moving operators

16. If Hamiltonian is invariant under inversion, then
    (a) J^2 and J_z commute
    (b) the space has intrinsic chirality or hardness
    (c) the parity operator is a constant of motion
    (d) the total angular momentum J commutes with the Hamiltonian

17. In the case of perturbation theory of non-degenerate stationary states with a perturbation $\lambda H^{(1)}$, the lowest order change in the energy $E_n^{(1)}$ of the nth state is

    (a) $\dfrac{\langle \psi_n^{(1)} H^{(0)} \psi_n^{(1)} \rangle}{\langle \psi_n^{(1)} \, \psi_n^{(1)} \rangle}$

    (b) $\dfrac{\langle \psi_n^{(1)} H^{(1)} \psi_n^{(0)} \rangle}{\langle \psi_n^{(1)} \, \psi_n^{(0)} \rangle}$

    (c) $\dfrac{\langle \psi_n^{(1)} H^{(1)} \psi_n^{(1)} \rangle}{\langle \psi_n^{(1)} \, \psi_n^{(1)} \rangle}$

    (d) $\dfrac{\langle \psi_n^{(0)} H^{(1)} \psi_n^{(0)} \rangle}{\langle \psi_n^{(0)} \, \psi_n^{(0)} \rangle}$

    where, $\psi_n^{(0)}$ and $\psi_n^{(1)}$ corresponds respectively to the unperturbed system and the 1st order correction to ψ in the perturbed system.

18. The WKB approximation is not valid
    (a) when h is large compared to the action
    (b) near a classical turning point
    (c) when the wavelength is very large
    (d) when the mass and energy are small

19. If ψ_S and ψ_A represented the symmetric and antisymmetric wave functions of the hydrogen molecule, then a bound state of two hydrogen atoms occurs
    (a) in the mixed symmetry state
    (b) in the symmetry state
    (c) in the asymmetry state
    (d) when the exchange integral vanishes

20. The dipole selection rules are
    (a) $\Delta l = 0, \pm 1, \Delta m = 0$
    (b) $\Delta l = 0, \pm 1, \Delta m = 0, \pm 1$
    (c) $\Delta l = \pm 1, \Delta m = 0, \pm 1$
    (d) $\Delta l = \pm 1, \Delta m = \pm 1$

21. The scattering amplitude is given by the formula

    $$f_K(\vec{K'}) = \frac{\sqrt{2\pi\mu}}{\hbar^2} \int d\tau' e^{i\vec{K'}\cdot\vec{r'}} V(r') \psi_K^{(+)}(\vec{r'})$$

    The first Born approximation consists of in
    (a) replacing $V(r)$ by $\frac{4\pi a}{m}$
    (b) $\psi_K^{(+)}(\vec{r})$ by a constant
    (c) putting $f_K(\vec{K'}) = 0$
    (d) replacing the exact eigen function $\psi_K^+(r')$ by a plane wave

22. The rapid oscillatory motion known as "Zitter-bewengung" of the Dirac-particle is
    (a) due to spin orbit coupling
    (b) due to existence of negative energy states
    (c) due to the fact that L_z is not concerned
    (d) because of relativistic invariance

23. The ground state wave function of the linear harmonic oscillator
    (a) is an exponentially decaying function
    (b) is a minimum uncertainty state
    (c) has odd parity
    (d) is negative for negative x-values

24. If two operators A and B satisfy the commutation relation $[A, B] = i\hbar$ then
    (a) all eigen states of A are eigen states of B
    (b) we can always find the common complete set of eigen states of A and B
    (c) $(\Delta A)(\Delta B) \geq \frac{1}{2}\hbar$, where (ΔA) and (ΔB) are uncertainties in A and B
    (d) $\left[\dfrac{A^2}{B}\right] = i\hbar$

25. Consider the coupled states of spin that can be formed from two spin $-\frac{1}{2}$ particles
   (a) the spin −1 coupled state is antisymmetric under interchange of the two spin $-\frac{1}{2}$ states
   (b) the spin −0 coupled state is symmetric under interchanged of the two spin $-\frac{1}{2}$ states
   (c) the spin −1 coupled state is symmetric under interchange of the two spin $-\frac{1}{2}$ states.
   (d) neither the spin −1 or the spin −0 coupled state has definite symmetry properties

26. Consider the orbital angular momentum operators L_x, L_y and L_z. Here
   (a) L^2 commutes with L_x, L_y and L_z where $L^2 = L_x^2 + L_y^2 + L_z^2$
   (b) it is not possible to find even one common eigen state of L_x, L_y and L_z
   (c) L_x, L_y and L_z can be simultaneously diagonalized
   (d) L_x, L_y and L_z can each be represented by 2-dimensional matrices

27. Corresponding to the hydrogen atom
   (a) the energy spectrum is discrete and equally spaced
   (b) the ground state is degenerate
   (c) the ground state is a minimum uncertainty state
   (d) the excited state are degenerate

28. Consider the Pauli's matrices $\sigma_x = \begin{pmatrix} 0 & 1 \\ 1 & 0 \end{pmatrix}$, $\sigma_y = \begin{pmatrix} 0 & -i \\ i & 0 \end{pmatrix}$ and $\sigma_z = \begin{pmatrix} 1 & 0 \\ 0 & -1 \end{pmatrix}$. Here

   (a) all the three matrices are simultaneously diagonalized
   (b) $\hbar \begin{bmatrix} 1 & 0 \\ 0 & -1 \end{bmatrix}$ is a matrix representation of S_z

(c) any 2 × 2 matrix can be written as the linear combination of the three Pauli's matrices

(d) $[\sigma_x, \sigma_y] = i\sigma_z$ (cyclic)

29. Let ψ be an eigen state of A with eigen value a. Let A commute with parity operator p, then

(a) $AP\psi$ is an eigen state of A with eigen value a
(b) every eigen state of A is an eigen state of P
(c) it is not possible to find a complete set of common eigen state
(d) $[A^2, P] \neq 0$

30. The quantity $\dfrac{\hbar^2}{me^2}$ has dimension of

(a) length
(b) inverse length
(c) energy
(d) inverse energy

31. In spherical polar coordinates

(a) the parity operation is defined by $r \to -r$ keeping θ and ϕ unchanged
(b) the parity operation is defined by $(r \to r), (\theta \to \pi - \phi)$ and $(\phi \to \pi + \phi)$
(c) the spherical harmonics $Y_{lm}(\theta, \phi)$ have even parity for all values of I
(d) $Y_{lm}(\theta, \phi)$ do not have definite parity for any value of I

32. Consider the one-dimensional free particle Schrodinger equation, Here

(a) $\psi(x,t) = e^{-i(kx-\omega t)}$ is a solution of the equation for a fixed k and ω
(b) $\psi(x,t) = e^{(x+ct)}$ is a solution of the equation for a fixed c
(c) $\psi(x,t) = \int a(k) e^{i(kx-\omega t)} dk$ is not a solution of the equation
(d) $\psi(x,t) = \log(x+ct)$ is a solution of the equation for a fixed c

33. The ground state wave function of a linear harmonic oscillator

(a) has exactly one node
(b) is a Gaussian in momentum space
(c) is an eigen state of position operator
(d) has odd parity

34. The energy levels of the bound electron in the hydrogen atom
    (a) are degenerate
    (b) are equally spaced
    (c) have no lower bound
    (d) have no upper bound

35. If two operators A and B commute with each other, we may conclude that
    (a) all eigen states of A are also eigen states of B
    (b) A and B are Hermitian operators
    (c) A^2 and B may or may not commute with each other
    (d) A and B have a complete set of common eigen states

36. A stationary state $\psi(r, t)$ of a quantum mechanical system
    (a) is a solution of the time-independent Schrodinger equation
    (b) satisfies $|\psi(r,t)|^2 = 1$ at every space point
    (c) is an eigen state of the momentum operator in general
    (d) is a minimum uncertainty state

37. If the potential $V(x)$ in a one-dimensional quantum mechanical system satisfies $V(x) = V(-x)$, all the energy eigen states of the system
    (a) have even parity
    (b) have odd parity
    (c) are not necessarily eigen states of the parity operator
    (d) have definite parity

38. The spherical harmonics $Y_{lm}(\theta, \phi)$
    (a) are eigen functions of $\vec{L} = L_x \hat{i} + L_y \hat{j} + L_z \hat{k}$ for all l and m
    (b) have eigen parity for all l
    (c) are angular wave functions of a particle moving in a central potential
    (d) cannot be separated for all l and m values in the form $f(\theta)g(\phi)$

39. If A represents a Hermitian operator
    (a) all its eigen values are real
    (b) its eigen values cannot be degenerate
    (c) its eigen values are either real or complex conjugates
    (d) its eigen values cannot be negative

40. Consider a system of two identical fermions. Here
    (a) total spin quantum number of the system must be zero
    (b) total wave function of the system must be symmetric under the interchange of particles
    (c) the probability of finding the two particles in the same one-particle state is zero
    (d) none of these

41. Consider the state of linear harmonic oscillator with eigen value $En = \left(n + \dfrac{1}{2}\right)\hbar\omega$. Let x and p represent the position and momentum operators of the oscillators. Then,
    (a) $\langle x \rangle = 0$ in this state
    (b) $\langle P \rangle > 0$ in this state
    (c) the uncertainty $(\Delta x) = 0$ in this state
    (d) the uncertainty $\Delta P = 0$ in this state

42. Let J_x, J_y, J_z be the components of angular momentum operator
    (a) J_x can be diagonalized, but not J_x and J_y
    (b) $[J^2, J_x + J_y + J_z] = 0$
    (c) $[J^2, J^2 - J_y^2 - J_x^2] \neq 0$
    (d) $[J_x, iJ_y, J_x - iJ_y] = 0$

43. The Dirac delta function satisfies one of the following properties.
    (a) $\delta_{(ax)} = |a^{-1}|\delta_{(x)}$
    (b) $\delta_{(ax)} = a\delta_{(x)}$
    (c) $\delta_{(ax)} = \delta_{(a)} \delta_{(x)}$
    (d) $\delta_{(ax)} = \delta_{(a)} + \delta_{(x)}$

44. The minimum possible uncertainty $\Delta x \cdot \Delta p = \dfrac{1}{2}\hbar$ is obtained for
    (a) $u_{(x)} = A\cos\dfrac{n\pi x}{2a}$ ($n \cdot$ odd)
    (b) $u_{(x)} = A\sin\dfrac{n\pi x}{2a}$ ($n \cdot$ even)
    (c) any harmonic oscillator wave function
    (d) $u_{(x)} = \left(\dfrac{2}{\pi}\right)^{\frac{1}{2}} \exp\left(-\dfrac{1}{2}\alpha^2 x^2\right)$

45. If u is a unitary matrix and $uz = az$, then
   (a) $a = 1$
   (b) $a > 1$
   (c) $a = \exp(\pm i\theta)$
   (d) $a = 0$

46. Let $\psi_i(n)\left(i = \alpha, \beta; n = \dfrac{1}{2}\right)$ be single-electron wave functions where α, β denote the state indices. Then the wave function of the two-electron system is
   (a) $\psi_\alpha(1)\psi_\beta(2)$
   (b) $\psi_\alpha(1) - \psi_\beta(2)$
   (c) $\psi_\alpha(1)\psi_\beta(2) + \psi_\alpha(2)\psi_\beta(1)$
   (d) $\psi_\alpha(1)\psi_\beta(2) - \psi_\beta(1)\psi_\alpha(2)$

47. The energy levels for a spherically symmetric potential are degenerate at least with respect to the quantum number m for $l > 0$. Then eigen functions $Y_{lm}(\theta, \phi)$ have the parity
   (a) $(-1)^m$
   (b) $(-1)^l$
   (c) $(-1)^{l-|m|}$
   (d) $(-1)^{l+|m|}$

48. A necessary condition for two observables A and B to be simultaneously measurable is that
   (a) the commutator $[A, B] = 0$
   (b) $A = B^\dagger$
   (c) the commutator $[A, B]$ is the unit operator
   (d) none of these

49. The lowest rank non-trivial matrix representation of the generators of the rotation group are

   (a) $\dfrac{1}{2}\begin{bmatrix} 0 & 1 \\ 1 & 0 \end{bmatrix}, \dfrac{1}{2}\begin{bmatrix} 0 & -i \\ i & 0 \end{bmatrix}, \dfrac{1}{2}\begin{bmatrix} 1 & 0 \\ 0 & -1 \end{bmatrix}$

   (b) $\dfrac{1}{2}\begin{bmatrix} 1 & 0 \\ 0 & 1 \end{bmatrix}, \dfrac{1}{2}\begin{bmatrix} -i & 0 \\ 0 & i \end{bmatrix}, \dfrac{1}{2}\begin{bmatrix} 0 & 1 \\ i & 0 \end{bmatrix}$

   (c) $\dfrac{1}{2}\begin{bmatrix} 0 & 1 \\ i & 0 \end{bmatrix}, \dfrac{1}{2}\begin{bmatrix} 0 & -i \\ i & 0 \end{bmatrix}, \dfrac{1}{2}\begin{bmatrix} 1 & 0 \\ 0 & -1 \end{bmatrix}$

   (d) $\dfrac{1}{2}\begin{bmatrix} 0 & 1 \\ 1 & 0 \end{bmatrix}, \dfrac{1}{2}\begin{bmatrix} 0 & -1 \\ 1 & 0 \end{bmatrix}, \dfrac{1}{2}\begin{bmatrix} 1 & 0 \\ 0 & -1 \end{bmatrix}$

50. Consider the wave packet $\psi(x,t) = \int_{-\infty}^{\infty} f(k)e^{i(kx-\omega t)} dk$. The phase velocity is given by

   (a) $\dfrac{\omega(k)}{k}$ at $k=0$

   (b) $\dfrac{dx}{dt}$ at $k=0$

   (c) $\dfrac{d\omega}{dk}$ at $k=0$

   (d) $\dfrac{d\omega}{dt}$ at $k=0$

51. Consider two spin $\dfrac{1}{2}$ states. The possible states in the $|j_1 m_1, j_2 m_2 >$ representation are

   (a) $\left|\dfrac{1}{2},\dfrac{1}{2}>\right|\dfrac{1}{2},\dfrac{1}{2}>\left|\dfrac{1}{2},\dfrac{1}{2}>\right|\dfrac{1}{2},-\dfrac{1}{2}>$
   $\left|\dfrac{1}{2},-\dfrac{1}{2}>\right|\dfrac{1}{2},\dfrac{1}{2}>\left|\dfrac{1}{2},-\dfrac{1}{2}>\right|\dfrac{1}{2},-\dfrac{1}{2}>$

   (b) $|1,0>|1,0>|1,0>|1,1>|1,0>|1,-1>$
   $|1,1>|1,-1>|1,1>|1,1>|1,-1>|1,-1>$

   (c) $\left|\dfrac{1}{2},\dfrac{1}{2}>\right|\dfrac{1}{2},-\dfrac{1}{2}>\left|-\dfrac{1}{2},-\dfrac{1}{2}>\right|\dfrac{1}{2},\dfrac{1}{2}>$
   $\left|\dfrac{1}{2},\dfrac{1}{2}>\right|\dfrac{1}{2},\dfrac{1}{2}>\left|\dfrac{1}{2},-\dfrac{1}{2}>\right|\dfrac{1}{2},-\dfrac{1}{2}>$

   $|0,0>\left|\dfrac{1}{2},\dfrac{1}{2}>|0,0>\right|\dfrac{1}{2},-\dfrac{1}{2}>|0,0>$

   (d) $\left|-\dfrac{1}{2},-\dfrac{1}{2}>|0,0>\right|-\dfrac{1}{2},\dfrac{1}{2}>$

52. Let J_x, J_y, J_z be the components of the angular momentum, then

   (a) J_x and J_y can be simultaneously diagonalized

   (b) $J^2 = J_x^2 + J_y^2 + J_z^2$ and J_z can be simultaneously diagonalized

(c) $J_x^2 + J_y^2 - J_z^2$ and J_x can be simultaneously diagonalized

(d) $J_x + iJ_y$ and $J_x - iJ_y$ can be simultaneously diagonalized

53. Let $|\psi>$ and $|x>$ be kets in a Hilbert space, then

   (a) $<x|\psi>$ is the wave function $<\psi|x>$ is the x representation

   (b) $<\psi|\psi> = \int \psi^*(x)\psi(x)dx$

   (c) $\psi(x)$ is an infinite dimensional ket vector in a Hilbert space

   (d) $\psi(x)$ is an infinite dimensional Bra vector in a Hilbert space

54. Let $\hat{x}$ and $\hat{p}$ be the 1-dimensional position and momentum operators respectively

   (a) in the momentum representation $\hat{x}$ is represented by $+i\hbar \dfrac{\partial}{\partial p}$

   (b) in the momentum representation $\hat{x}$ is a mere number

   (c) in the momentum representation $\hat{p} = \dfrac{\partial}{\partial p}$

   (d) in the position representation $\hat{p}$ is a mere number

55. Considering the Schrodinger and Heisenberg pictures in quantum mechanics.

   (a) In the Schrodinger picture, operator cannot have any explicit time-dependence

   (b) At any time, expectation value of physical observables are equal in Schrodinger and Heisenberg pictures.

   (c) Both state vectors and operators evolve as per appropriate evolution equations in the Heisenberg pictures.

   (d) The position operator does not evolve with time in the Heisenberg picture.

56. Consider the gamma matrices $\gamma^0, \gamma^1, \gamma^2$ and γ^3. If ψ is the Dirac wave function and $\bar{\psi} = \psi^+ \gamma^0$

   (a) $\psi^1 \gamma^0 \psi$ transforms as a component of Lorentz 4-vector

   (b) each γ matrix has trace equal to 4

   (c) $\left[\gamma^\mu \gamma^\upsilon\right] = 0$, $\mu\upsilon = 0, 1, 2, 3$

   (d) $i\bar{\psi}\gamma^0\gamma^1\gamma^2\gamma^3\psi$ transforms like $\bar{\psi}\psi$ under Lorentz transformations

57. Consider the Klein–Gordon equation. If ϕ is the wave function
    (a) the conserved charge density is given by $\phi^* \dfrac{\partial \phi}{\partial t}$ where t is the time
    (b) the conserved current density is proportional to $(\phi^* \nabla \phi - \phi \nabla \phi^*)$
    (c) the equation represents the spin 0 and spin $\dfrac{1}{2}$ particles
    (d) the equation of the 1st order in the space derivative $\overline{\nabla}$

58. Consider a unitary operator u and a wave function ψ
    (a) $u\psi = e^{i\theta l z}$ corresponds to rotating ψ about Z-axis by an angle θ if l_z is the orbital angular momentum operator $(xPy - yPx)$
    (b) the rotation transformation on ψ cannot be implemented by any unitary operator u
    (c) $u\psi$ has the same norm as ψ
    (d) u can be represented in general by the 2 × 2 matrix
    $\begin{bmatrix} \cos\theta & \sin\theta \\ -\sin\theta & \cos\theta \end{bmatrix}$

59. If the three vectors $\overline{a}, \overline{b}$ and $\overline{c}$ are linearly independent,
    (a) $\alpha \overline{a} + \beta \overline{b} + \gamma \overline{c} = 0 \Rightarrow \alpha$ or β or γ is 0
    (b) none of the three vectors is the null vector
    (c) two or three vectors are necessarily orthogonal to each other
    (d) $\overline{a}, \overline{b}$ and $\overline{c}$ form a basis in a 2-dimensional space

60. Consider the Born approximation
    (a) scattering amplitude considered as a function of the momentum transferred to the particle when it encounters a potential, is the Fourier transform of the potential apart from constant factors
    (b) the scattering amplitude explicitly depends on the initial momentum
    (c) the scattering amplitude is real for any form of the potential
    (d) the Born approximation holds for wave function deviating by a large value from the free particle wave function

61. The inner product of two vectors ψ and ϕ in a linear vector space satisfies
    (a) $(\psi, a\phi) = a^*(\psi, \phi)$ where a is the complex number
    (b) $(\psi, \phi) = (\phi, \psi)$

(c) $(\psi, \psi) = 0$ if ψ is a null vector

(d) $|(\psi, \phi)| > \|\psi\| \|\phi\|$ for any ψ and ϕ

62. The dipole selection rules are

(a) $\Delta l = 0, \Delta m = 0$

(b) $\Delta l = 2, \Delta m = 1$

(c) $\Delta l = \pm 1, \Delta m = 0 \pm 1$

(d) $\Delta l = 2, \Delta m = 2$

63. A system is known to be in a state described by the wave function

$$\psi(\theta, \phi) = \frac{1}{\sqrt{30}}\left(5Y_4^0 + Y_6^0 - 2Y_6^3\right)$$

where $Y_l^m(\theta, \phi)$ are spherical harmonics. The probability of finding the system in a state with $m = 0$ is

(a) 0 (b) $\frac{2}{15}$ (c) $\frac{1}{4}$ (d) $\frac{13}{15}$

64. A fast charged particle passes perpendicularly through a thin glass sheet of refractive index 1.5. Inside the glass, the particles emit light. The minimum speed of the particle is

(a) $\frac{C}{3}$ (b) $\frac{2C}{3}$ (c) $\frac{4C}{9}$ (d) $\frac{C}{9}$

65. The degeneracy of the first excited state of a 3-dimensional harmonic oscillator is

(a) 1 (b) 2 (c) 3 (d) 6

66. The energy of a single particle in an infinite potential box is $E_o n^2$ where $E_o = \frac{\hbar^2 \pi^2}{2ma^2}$, $n = 1, 2,...$ If three non-interacting bosons are in such a potential box, the ground state energy of the system will be

(a) $3E_o$ (b) $4E_o$ (c) $6E_o$ (d) $9E_o$

67. The momentum of an electron (mass m) which has the same KE as its rest mass energy (c is velocity of light) is

(a) $\sqrt{3}\,mc$ (b) $\sqrt{2}\,mc$ (c) mc (d) $\frac{mc}{\sqrt{2}}$

68. The eigen function for the Hamiltonian of a particle of mass m in one-dimensional potential $V(x)$ is given to be $\psi(x) = A \exp\left(\dfrac{bx^2}{2}\right)$ where A and b are constants. It follows that
   (a) $V(x) =$ constant
   (b) $V(x) \propto x^3$
   (c) $V(x) \propto x^2$
   (d) $V(x) \propto \dfrac{1}{x}$

69. A free particle in one dimension is in a state described by $\psi(x, t) = \dfrac{A\exp(ipx - iEt)}{\hbar} + \dfrac{B\exp(-ipx - iEt)}{\hbar}$. The probability current density is given by
   (a) $\dfrac{p}{m}$
   (b) $(|A| - |B|)\dfrac{p}{m}$
   (c) $\left(|A|^2 - |B|^2\right)\dfrac{p}{m}$
   (d) $\left(|A|^2 + |B|^2\right)\dfrac{p}{m}$

70. Kinetic energy of a relativistic particle of rest m moving with a speed v is
   (a) $\dfrac{1}{2}mv^2$
   (b) $\dfrac{mc^2}{\sqrt{1 - \dfrac{v^2}{c^2}}}$
   (c) $\dfrac{mc^2}{\sqrt{1 - \dfrac{v^2}{c^2}}} - mc^2$
   (d) $\dfrac{1}{2m(v^2 - c^2)}$

71. In the first excited state of a one-dimensional harmonic oscillator with angular frequency ω, the energy eigen value is given by
   (a) $\left(\dfrac{1}{2}\right)\hbar\omega$
   (b) $\hbar\omega$
   (c) $\left(\dfrac{3}{2}\right)\hbar\omega$
   (d) $2\hbar\omega$

72. Degeneracy of the first excited state of an isolated hydrogen atom is
   (a) 2
   (b) 4
   (c) 6
   (d) 8

73. Probability current density in quantum mechanics is expressed by the operator

(a) $\dfrac{i\hbar}{2m}\left(\psi\vec{\nabla}\psi^* - \psi^*\vec{\nabla}\psi\right)$
(b) $\dfrac{\hbar}{2m}\left(\psi\vec{\nabla}\psi^* - \psi^*\vec{\nabla}\psi\right)$
(c) $\dfrac{\hbar}{2m}\left(\psi\vec{\nabla}\psi^* + \psi^*\vec{\nabla}\psi\right)$
(d) $\dfrac{i\hbar}{2m}\left(\psi\vec{\nabla}\psi^* + \psi\vec{\nabla}\psi\right)$

74. If the ϕ dependent part of the eigen function of an electron in a hydrogen atom is $e^{2i\phi}$, the minimum principal and orbital quantum number n and l respectively for the eigen function will be

(a) $n = 2, l = 1$
(b) $n = 3, l = 2$
(c) $n = 4, l = 3$
(d) $n = 3, l = 1$

75. The Heisenberg uncertainty relation in the simultaneous determination of the position and momentum arises because

(a) the position only has uncertainty
(b) the momentum only has uncertainty
(c) the position and momentum operators commute
(d) the position and momentum operators do not commute

76. The eigen function of a harmonic oscillator are

(a) Hermite polynomials
(b) Langer polynomials
(c) Legendre polynomials
(d) associated Legendre polynomials

77. The allowed energy levels E_n for a particle of mass m in one-dimensional infinite potential are given by

(a) $E_n = \dfrac{\pi n^2 h^2}{ma}$
(b) $E_n = \dfrac{n^2 \hbar^2}{2ma^2}$
(c) $E_n = \dfrac{\pi^2 n^2 h^2}{2ma^2}$
(d) $E_n = \dfrac{\pi^2 nh^2}{2ma^2}$

78. The experiment which gives the evidence for the existence of spin is

(a) Davis and Germer
(b) Stern–Gerlach
(c) G.P. Thomson
(d) Michelson–Morley

79. The hyperfine structure of spectral lines can be explained based on
    (a) spin–orbit interaction
    (b) spin–spin interaction
    (c) electron–nuclear interaction
    (d) proton–neutron interaction

80. If ψ_m is an energy eigen function, the property represented by $\int \psi_m^* \psi_n d(\tau) = \delta'_{mn}$ where δ is Kronecker delta is
    (a) orthonormality
    (b) orthogonality
    (c) normalization
    (d) degeneracy

81. For a particle of mass m in a one-dimensional square well of width $2a$, the energy difference between the two lowest levels is
    (a) $\dfrac{\pi^2 h^2}{8ma^2}$
    (b) $\dfrac{2\pi^2 h^2}{8ma^2}$
    (c) $\dfrac{3\pi^2 h^2}{8ma^2}$
    (d) $\dfrac{4\pi h^2}{8ma^2}$

82. The radial part of the waveform for the hydrogen atom is expressed in terms of
    (a) Laguerre polynomial
    (b) Hermite polynomial
    (c) Legendre polynomial
    (d) associated Laguerre polynomial

83. Which of the following is not Hermitian
    (a) $i\nabla$
    (b) ∇^2
    (c) x
    (d) ∇

84. For scattering by a rigid sphere, quantum mechanics gives a value for the cross section which is m times the classical value. The value of m is
    (a) 1
    (b) 2
    (c) 3
    (d) 4

85. In the Stern–Gerlach experiment, the number of components in which the atomic beam splits depends upon the value of
    (a) l
    (b) s
    (c) j
    (d) g

86. If n is the principal quantum number for a hydrogen electron, the spin-orbit interaction energy is proportional to
    (a) n^2
    (b) n^4
    (c) n^{-5}
    (d) n^5

87. If the wave function of a particle is $\exp(i\,\vec{k},\vec{r})$ then the eigen value of the momentum operator is

(a) $i\vec{k}$ (b) $\dfrac{h\vec{k}}{2\pi}$ (c) $\pi\vec{k}$ (d) $\dfrac{k^2}{2}$

88. The expectation value of the position operator for a simple harmonic oscillator in its nth state is

(a) zero if n is even
(b) zero if n is odd
(c) zero for all values of n
(d) non-zero for all values of n

89. For an electron in the d-state, the possible values of the angular momentum j and the projection m are given by

(a) $j = \dfrac{5}{2}, m = 0$

(b) $j = \dfrac{3}{2}, m = -\dfrac{1}{2}$

(c) $j = 2, m = -\dfrac{1}{2}$

(d) $j = \dfrac{1}{2}, m = -\dfrac{1}{2}$

90. The ground state wave function of the hydrogen atom is proportional to (a is a positive constant)

(a) $\exp\left(\dfrac{-r^2}{a^2}\right)$ (b) $\exp\left(\dfrac{r}{a}\right)$

(c) $\exp\left(\dfrac{-a}{r}\right)$ (d) $\exp\left(\dfrac{-r}{a}\right)$

91. Which of the following relations is satisfied by $W = \langle f|H|f\rangle$ where it is the Hamiltonian, δ is a trial wave function, and E_g is the ground state energy

(a) $W < E_g$ (b) $W \leq E_g$ (c) $W \geq E_g$ (d) $W > E_g$

92. Choose the wrong statement concerning scattering.

(a) The Born approximation obeys the optical theorem.
(b) The Born approximation violates the optical theorem.

(c) There are interference terms in the differential cross section in partial wave analysis.

(d) There are no interference terms in the total cross section in the partial wave analysis.

93. Which of the following quantities is not quantized in quantum theory?

(a) angular momentum of a photon
(b) charge of an elementary particle
(c) energy of a linear harmonic oscillator
(d) linear momentum of a free particle

94. The probability current density cannot be experimentally measured as such measurement is inconsistent with

(a) uncertainty principle
(b) correspondence principle
(c) exclusion principle
(d) principle of least action

95. In conventional notation, the eigen function of a linear harmonic oscillator is of the form $\psi = NF(\alpha x)e^{-\frac{1}{2}\alpha^2 x^2}$, where N is a constant and the function $F(\alpha x)$ is a

(a) Bessel function
(b) Legendre function
(c) Laguerre function
(d) Hermite function

96. The first radial function of a hydrogen-like atom is

(a) $\left(\dfrac{Z}{a_0}\right)^{\frac{3}{2}} 2e^{-\frac{2r}{a_0}}$

(b) $\left(\dfrac{2z}{a_0}\right)^{\frac{1}{2}} e^{-\frac{2r}{a_0}}$

(c) $\left(\dfrac{Z}{a_0}\right)^{\frac{3}{2}} e^{-\frac{2zr}{a_0}}$

(d) $\left(\dfrac{Z}{a_0}\right)^{\frac{3}{2}} e^{-\frac{2r}{2a_0}}$

97. The wave function of a free particle constrained to move in one dimension is $\psi(x) = a \exp-\left(ihx - \dfrac{x^2}{2A}\right)$, the expectation value of momentum is

(a) hK

(b) $H\left(K + \dfrac{1}{\wedge}\right)$

(c) $h\left(K + \sqrt{\wedge}\right)$

(d) $h\left(K - \dfrac{1}{\wedge}\right)$

98. A particle of energy E and mass m moving in one dimension is incident on a potential step as shown. The wave function in this region x_0 for E_0 is proportional to

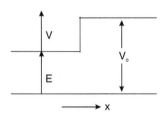

(a) $\exp\left(-i\sqrt{2m(V_0 - E)}\frac{x}{h}\right)$
(b) $\exp\left(-\sqrt{2m(V_0 - E)}\frac{x}{h}\right)$

(c) $\sinh\left((V_0 - E)\frac{x}{h}\right)$
(d) $\sin\left[\sqrt{2m(V_0 - E)}\frac{x}{h}\right]$

99. The value of commutation relation $[P, x^3]$ is

(a) $-3ihx^2$
(b) zero

(c) $3[x^2 p - px^3]$
(d) none of the above

100. The Hamiltonian of quantum system is given by
$$H = \begin{bmatrix} E_0 & 0 & 0 \\ 0 & E_1 & b \\ 0 & b & E_1 \end{bmatrix},$$ what are its eigen values?

(a) $E_0 = E_1 + b$
(b) $E_1 \pm b$

(c) $E_0, E + b, E - b$
(d) $E_0 \pm b, E_1$

101. A quantum particle of mass m confined to a two-dimensional harmonic potential $V = \frac{1}{2}\omega^2(x^2 + y^2)$ has the energy eigen values E_1 with degeneracy d (n and I are integers), then

(a) $E = h\omega(n+1); d = n+1$
(b) $E = h\omega\left(n+\frac{1}{2}\right); d = 1$

(c) $E = h\omega\left(n+\frac{1}{2}\right); d = 2I + 1$
(d) $E = nh\omega; d = 1$

102. An electron in an atom has orbital angular momentum $2h$. What are the possible values of its total angular momentum?

(a) $3h, h$
(b) $\frac{5}{2}h, \frac{3}{2}h$
(c) $\frac{3}{2}h, \frac{1}{2}h$
(d) $\pm\frac{3}{2}h$

103. The time dependence of a quantum state ψ_n with eigen energy E_n is given by

    (a) $\exp\left[\dfrac{-iE_n t}{h}\right]$
    (b) $\exp\left[\dfrac{iE_n t}{h}\right]$
    (c) $\sin\left[\dfrac{E_n t}{h}\right]$
    (d) $\exp\left[\dfrac{-ht^2}{E_n}\right]$

104. Which of the following statements is correct about Born approximation for calculating electron atom scattering cross section?
    (a) It is valid for very low energy electrons.
    (b) It is valid for fast electrons but not at relativistic energies.
    (c) It is good for any electron energy, provided atomic charge is large.
    (d) It is valid for medium energy.

105. The electron eigen function $\psi_n, l, m\, (r, \theta, \phi)$ in a hydrogen atom with quantum numbers n, l and m varies at small r as
    (a) r^n
    (b) r'
    (c) $r^n + 1$
    (d) $\dfrac{1}{r}$

106. Which of the following statements about spin–orbit coupling is correct?
    (a) It is weak as it has relativistic effect.
    (b) It arises due to mutal coulomb repulsion between the electrons of the atoms.
    (c) It arises due to exchange symmetry.
    (d) It arises because nucleus is not a point charge.

107. The wave function ψ_n for a particle in an infinite well of width L is $\psi_n = A \sin\left(n\pi \cdot \dfrac{X}{L}\right)$. The normalization constant A
    (a) depends on n only
    (b) depends on L only
    (c) depends on n and L
    (d) is independent of n and L

108. The variation method yields
    (a) wave function for ground state
    (b) exact value of ground state

(c) upper limit to ground state energy
(d) value of excited state energy

109. In the coordinate representation, one possible relation satisfied by K, the generator of infinitesimal transformations is
   (a) $\langle x''|K|x'\rangle = i\delta(x-x')$
   (b) $\langle x''|K|x'\rangle = i\left(\dfrac{\partial}{\partial x'} - \dfrac{\partial}{\partial x''}\right)$
   (c) $\langle x'|K|x''\rangle = 0$
   (d) $\langle x''|K|x'\rangle = i\left(\dfrac{\partial}{\partial x'}\right)\delta(x'-x'')$

110. Consider the translational operator $D\xi : D\xi|x\rangle = |x+\xi\rangle$ if we represent $D\xi = (\exp iA(\xi))$ then
   (a) $A(\xi)$ is unitary
   (b) $A(\xi)$ is a projection operator
   (c) $A(\xi)$ is Hermitian
   (d) $A(\xi)$ is antiunitary

111. The equation of motion in the Heisenberg picture is given as
   (a) $H\psi = E\psi$
   (b) $\psi(t) = e^{\left[\frac{-i}{\hbar}Ht\right]}\psi(0)$
   (c) $i\hbar \dot{A} = [A, H]$
   (d) $i\hbar \dfrac{\partial \psi}{\partial t} = H\psi$

112. The presence of a uniform electric field
   (a) leads to conservation parity
   (b) leads to commutativity of rotations and space inversions
   (c) causes states of opposite parity to be mixed
   (d) makes parity to commute with Hamiltonian

113. In an anharmonic oscillator with a perturbation of the form λx^4 the matrix element $\langle x'|x^4|n\rangle$ evaluated between unperturbed states $|n\rangle$ satisfies the relation
   (a) $\langle n'|x^4|n\rangle = 0$ values $n'-n = \pm 1, \pm 3...$
   (b) $\langle n'|x^4|n\rangle = 0$ values $n'-n = 0, \pm 2, \pm 4...$
   (c) $\langle n'|x^4|n\rangle = A\delta n, n'$
   (d) $\langle n'|x^4|n\rangle \neq 0$ only if x^4 commutes with it

114. WKB approximation
   (a) overestimates the energy levels
   (b) underestimates the energy levels
   (c) gives the energy levels of the linear harmonic oscillator correctly
   (d) gives only the first two levels of the linear harmonic oscillator

115. Consider the transition matrix element
$$Me = \left\langle K \left| \sum_{i=1}^{Z} r_i^e P_l\left(\hat{r}_i - \hat{K}(t')\right) \right| S \right\rangle$$
   where, Z is the total charge on the target. The value of M_0 is
   (a) $\sin\left(\hat{r}_i \cdot K\right)$
   (b) $\cos\left(\hat{r}_i \cdot K\right)$
   (c) 0
   (d) depends on $\langle K|$ and $|S\rangle$

116. The contribution to the total cross section σ from the lth partial wave is
   (a) $\dfrac{4\pi}{K^2}(2l+1)\sin^2 \delta_l$
   (b) $\dfrac{4\pi}{RI_m f(0)}$
   (c) $\dfrac{(x+1)}{R}e^{i\delta_l}\sin\delta_l P_l(\cos\theta)$
   (d) $\left|\dfrac{\sin\left(kr - \dfrac{l\pi}{2} + \delta_l\right)}{kr}\right|^2$

117. Consider the scattering of two particles of masses m_1 and m_2. The scattering angle θ in the cm frame and the corresponding angle θ_0 in the laboratory frame are related by
   (a) $\tan\theta_0 = m_1 \sin\theta + m_2 \cos\dfrac{\theta}{m_1 + m_2}$
   (b) $\tan\theta_0 = \dfrac{\sin\theta}{\cos\theta + \dfrac{m_1}{m_2}}$
   (c) $\tan\theta_0 = \dfrac{\cos\theta}{m_1 \sin\theta + m_2 \cos\theta}$
   (d) $\tan\theta_0 = \dfrac{\tan\theta_0}{m_1 + m_2 \sin\theta_0}$

118. The Klein–Gordon equation is unsatisfactory because
   (a) this equation does not have a Schrodinger-like form
   (b) the density ρ is not positive and the spectrum is not bounded below
   (c) the reflected flux is larger than the incident one
   (d) one cannot localize the particle within a distance of the order of the Compton wavelength

119. Let $\psi(x)$ be the normalized wave function corresponding to an infinite one-dimensional system
   (a) $\psi^*(x)\psi(x)$ is a conserved quantity
   (b) $\int_{-\infty}^{\infty} \psi^*(x)\psi(x)\,dx$ is conserved quantity
   (c) $\dfrac{i\hbar}{2m}\left(\psi^* \dfrac{\partial \psi}{\partial x} - \psi \dfrac{\partial \psi^*}{\partial x}\right)$ is conserved quantity
   (d) $\psi(x)$ and $\psi^*(x)$ are independently conserved

120. Consider the motion of a free particle in one dimension with energy E and with the definition $\alpha = \dfrac{2mE}{\hbar^2}$. Let x denote the position.
   (a) Any value E of the particle is doubly degenerate.
   (b) There is no degenerate energy eigen value for this system.
   (c) $e^{\alpha x}$ is an eigen state of energy.
   (d) Linear combination of $e^{\alpha x}$ and $e^{-\alpha x}$ are eigen states of energy.

121. Let A and B be two commuting operations with ϕ_a being an eigen state of A with eigen value a.
   (a) ϕ_a is also an eigen state of B in general.
   (b) ϕ_a is an eigen state of B only if a is non-degenerate eigen value.
   (c) $B\phi_a$ is not an eigen state of A.
   (d) These are not states in which ΔA and ΔB both vanish where ΔA is the uncertainty in A and ΔB that in B.

122. Consider a one-dimensional physical system with potential $V(x)$ which has even parity.
   (a) All energy eigen states of the system have definite parity.
   (b) All energy eigen states of the system have even parity.

(c) All the energy eigen states of the system have odd parity.

(d) The ground state of the system is degenerate.

123. Consider a linear harmonic oscillator with ground state wave function $\psi(x)$. Here

   (a) $\int_{-\infty}^{\infty} \psi(x) e^{ikx} dx$ is a Gaussian

   (b) the expectation values of the position operator with respect to $\psi(x)$ is unity

   (c) $\psi(x)$ has odd parity

   (d) the expectation value of the momentum operator with respect to $\psi(x)$ is unity

124. Consider a charged particle with linear momentum $\vec{p}$ in a uniform magnetic field. Here

   (a) the energy eigen spectrum is purely discrete and unequally spaced

   (b) the energy eigen spectrum is purely discrete and equally spaced

   (c) the vector potential $\vec{A}$ is unequally fixed for a given value of magnetic field

   (d) the Hamiltonian can be written as $H = \dfrac{\pi^2}{2m}$, where $\pi = \vec{p} - \dfrac{eA}{c}$

125. Let A, B, C be three operators.

   (a) If $[A, B] = 0$ and $[A, C] = 0$ then $[B, C]$ is also necessarily zero.

   (b) Any common eigen states of A and B is also an eigen state of C.

   (c) $[A^2, BC] = 0$.

   (d) There exists a complete set of common eigen states of A, B and C.

126. Let $\sigma_x, \sigma_y, \sigma_z$ be three Paulis matrices.

   (a) All the three matrices can be simultaneously diagonalized.

   (b) All the three matrices can be chosen to have all elements real.

   (c) Any (2×2) matrix, can be expanded in terms of $\sigma_x, \sigma_y, \sigma_z$ and the (2×2) indentify matrix.

   (d) Two of the three σ matrices are Hermitian and the third is anti-Hermitian.

127. Consider the hydrogen atom
   (a) the ground state is a minimum uncertainty state
   (b) all energy eigen values are degenerate
   (c) the eigen spectrum is equally spaced
   (d) the energy spectrum has both discrete and continuous part

128. Let A and B be two Hermitian matrices, then
   (a) e^{iA} and e^{iB} are unitary matrices
   (b) $e^{i(A+B)} = e^{iA} \cdot e^{iB}$ in general
   (c) it is possible to always find the complete set of common eigen states of A and B
   (d) $\alpha A + \beta B$ is Hermitian for any two complex numbers α and β

129. If u is a unitary operator,
   (a) $(u\psi, u\phi)$ is a number different from (ψ, ϕ) for any two vectors ψ and ϕ
   (b) $u\phi_1, u\phi_2, ..., u\phi_k$ is an orthonormal basis if $\phi_1, \phi_2, ..., \phi_k$ is an orthonormal basis
   (c) $\|u\| = 0$
   (d) $\|u\| > 1$

130. Consider the Hamiltonian $\hat{H}$ of a linear harmonic oscillator.
   (a) $\hat{H}$ is bounded operator.
   (b) The spectrum $\hat{H}$ is partly discrete and partly continuous.
   (c) The operator $\exp(-\beta \hat{H})$ where β is a positive constant is an unbounded operator.
   (d) All the eigen values of the operator $\exp\left(\dfrac{i\hat{H}t}{\hbar}\right)$, where t is a real number, having unit moduli.

131. The transformation from the Schrodinger picture to Heisenberg picture
   (a) is an orthogonal transformation
   (b) is a unitary transformation
   (c) is possible only if the Hamiltonian is explicitly time-dependent
   (d) is such that the commutator $[\hat{A}_H(t), \hat{B}_H(t)]$ between any two arbitrary opertors $\hat{A}$ and $\hat{B}$ is equal to $[\hat{A}_s, \hat{B}_s]$

132. To form a basis in Hilbert space, a set of vectors
    (a) must be linearly independent and must span the space
    (b) must span the space, but need not be linearly independent
    (c) must be linearly independent, but need not span the space
    (d) must be mutually orthogonal

133. The Klein–Gordon equation $(\pi + m^2)\phi = 0$
    (a) describes photons
    (b) describes electrons
    (c) is satisfied by each component of the Dirac wave function for spin $-\frac{1}{2}$ particle
    (d) is the non-relativistic equation corresponding to Dirac equation

134. Partial-wave analysis is useful in scattering theory
    (a) only if the potential is spherically symmetric
    (b) for any arbitrary potential
    (c) only if the potential vanishes beyond a certain distance
    (d) only for the coulomb potential

135. For a system with Hamiltonian $H = H_0 + H'(t)$ any operator $F(t)$ in the interaction picture satisfies the equation $i\hbar \frac{\partial E}{\partial T}(t)$
    (a) $[F(t), H_0]$
    (b) $F(t) H_0 F^{-1}(t)$
    (c) $H_0 F(t) + F(t) H_0$
    (d) $H_0^{-1} F(t) H_0$

136. The expression for the probability current density for a relativistic particle
    (a) is identical to the corresponding non-relativistic expression in all cases
    (b) depends on the spin of the particle
    (c) is invariant under Lorentz transformations
    (d) is always pure imaginary

137. In a constant external electric field the third excited state of the hydrogen atom displays
    (a) the linear Stark effect but not the quadratic Stark effect
    (b) both linear and quadratic Stark effects
    (c) only the quadratic and not the linear Stark effect
    (d) only the cubic Stark effect

138. The transformation from laboratory frame to the centre of mass frame
 (a) is possible for central potentials alone
 (b) is possible for short-ranged potentials alone
 (c) leaves the differential cross section unaltered
 (d) leaves the total cross section unaltered

139. Consider the wave packet $\psi(x,t) = \int a(k) e^{i(kx-\omega t)} dk$, where $a(k)$ is negligible except when k lies in a small internal Δk. Then the wave packet moves with a group velocity
 (a) $\dfrac{d\omega}{dk}$
 (b) $\dfrac{\omega}{R}$
 (c) $\dfrac{dx}{dt}$
 (d) $\sqrt{\dfrac{d^2x}{dt^2}}$

140. The width $\Delta x = (\langle x^2 \rangle - \langle x \rangle^2)^{\frac{1}{2}}$ of the Gaussian wave packet $\psi = N \exp\left(\dfrac{-x^2}{2\sigma^2}\right)$ where origin is chosen at the centre of the wave packet is
 (a) $\dfrac{\sigma}{\sqrt{2}}$
 (b) $\dfrac{\hbar}{\sqrt{2\sigma}}$
 (c) $\dfrac{\sigma^2}{2}$
 (d) 1

141. If p is a projection operator, then
 (a) $p^2 = p$
 (b) $p = I$
 (c) $p = p^{-1}$
 (d) $p^n = 0 (n > 1)$

142. The single state of a system composed of two particles of spin $\dfrac{1}{2}$ is represented by
 (a) $\left|\dfrac{1}{2}, \dfrac{1}{2}\right\rangle$
 (b) $\dfrac{1}{\sqrt{2}}\left\{\left|\dfrac{1}{2}, \dfrac{1}{2}\right\rangle + \left|-\dfrac{1}{2}, \dfrac{1}{2}\right\rangle\right\}$
 (c) $\left|-\dfrac{1}{2}, -\dfrac{1}{2}\right\rangle$
 (d) $\dfrac{1}{\sqrt{2}}\left\{\left|\dfrac{1}{2}, -\dfrac{1}{2}\right\rangle - \left|-\dfrac{1}{2}, \dfrac{1}{2}\right\rangle\right\}$

143. The value of the integral $\dfrac{1}{2\pi} \int_0^{2\pi} dy \int_{-\infty}^{\infty} du\, e^{iu(x-y)} \cos(x-y) e^{\frac{-(x-y)^2}{2}}$ is
 (a) 2π
 (b) e^{-x^2}
 (c) $\cos 2x$
 (d) 0

144. If [A, B] denotes the commutator of operators A and B then
[A, [B,C]] + [B,[C,A]] + [C,[A,B]] = 0 is true
(a) always
(b) if A, B and C are Hermitian
(c) if [A,B] = [B,C] = [C,A]
(d) if at least two of the operators are commuting

145. The average dipole moment $\langle \vec{d} \rangle$ of a system of N particles is given by $\langle \vec{d} \rangle = \int d\vec{r}_1 \int d\vec{r}_2 ... \int d\vec{r}_N \psi^*(\vec{r}_1 - \vec{r}_N) \cdot \left(\sum_{i=1}^{N} e_i \vec{r}_i \right) \psi(\vec{r}_1 ... \vec{r}_N)$

where, e_i is the charge on the ith particle with position vector $\vec{r}_i$. If the system is in a state of well defined parity then $\langle \vec{d} \rangle$ equals

(a) 1
(b) 0
(c) $N \sum_{i=1}^{N} e_i$
(d) N

146. Solution of the Schrodinger equation with a periodic potential of period T can be written as
(a) $\psi^{\pm}(x) = \exp(ikx \pm 2\pi T)$
(b) $\psi^{\pm}(x) = \exp\left(\frac{i}{k} \pm 2\pi/T\right)(x)$
(c) $\psi^{\pm}(x) = \exp\left(\frac{ik}{Tx}\right) u^{\pm}(x)$, where, $u^{\pm}(x+T) = u^{\pm}(x)$
(d) $\psi^{\pm}(x) = u^{\pm}(x)$, where, $u^{\pm}(x+T) = u^{\pm}(x)$

147. Let $\vec{\sigma} = (\sigma_x, \sigma_y, \sigma_z)$ where σ_x, σ_y and σ_z are Paulis matrices and $\hat{n}$ be a unit vector in an arbitrary direction, then the eigen values of $(\vec{\sigma}, \hat{n})$ are
(a) 0, 1
(b) ±1
(c) 1, doubly degenerate
(d) −1, doubly degenerate

148. The energy levels of a charged particle in a magnetic field of strength $\vec{B}$ are
(a) proportional to $B^2 \cdot n^2$
(b) given by $n\hbar^2 B^2$
(c) uniformly spaced
(d) proportional to $B^{\frac{1}{2}} \cdot n$

149. A particle of mass m moves in the region $0 \leq x < \infty$ in the potential
$$V(x) = \begin{cases} \frac{1}{2}m\omega^2 x^2 & \text{for } x > 0 \\ +\infty & \text{for } x = 0 \end{cases}$$
. The ground state energy is

(a) 0 (b) $\frac{1}{4}\hbar\omega$ (c) $\frac{1}{2}\hbar\omega$ (d) $\frac{3}{2}\hbar\omega$

150. The most appropriate nucleon potential for a deuteron is
   (a) a square well of finite depth
   (b) an infinitely deep square well
   (c) a harmonic oscillator potential
   (d) $V_{(A)} = -(\text{Const})r^{-2}$

151. Let L_x, L_y and L_z denote the components of the angular momentum of a classical rotating object. The Hamiltonian of the system is given by $H = \frac{L_x^2 + L_y^2}{2A}$ where A is a positive constant. Then
   (a) L_x is conserved
   (b) L_x, L_y and L_z are conserved
   (c) L_x and L_y are conserved but not L_z
   (d) $|L|$ is conserved

152. For a quantum mechanical particle moving in one dimension (in various potential), select the physically bound state wave functions (Here $-\infty < x < \infty$ and b, c are positive constants).

   (a) $\frac{c}{x} - b^2$

   (b) $c \exp\left(\frac{-|x|}{b}\right)$

   (c) $c \exp\left(\frac{-x^2}{b^2}\right)$

   (d) $\frac{c}{(x^2 + b^2)^{\frac{1}{4}}}$

153. Let the Hermitian operator H be the Hamiltonian of a quantum mechanical particle, E_0 its ground state energy and $\psi(\vec{r})$ an arbitrary normalized trial wave function. The quantity $\int \psi^*(\vec{r}) H \psi(\vec{r}) dV$
   (a) must necessarily be less than or equal to E_0
   (b) must necessarily be greater than or equal to E_0
   (c) may be any real number greater than, equal to or less than E_0
   (d) is purely imaginary

154. The Hamiltonian of the electron in the hydrogen atom is $H = \dfrac{P^2}{2m} - \dfrac{e^2}{(4\pi\varepsilon_0 r)}$. Let $|\psi\rangle$ denote the ground state of H, with $\langle \psi | \psi \rangle = 1$, then
   (a) $\langle \psi | r | \psi \rangle$ is zero
   (b) $|\psi\rangle$ is an eigen state of $\dfrac{P^2}{2m}$
   (c) $\langle \psi | \dfrac{1}{r} | \psi \rangle$ is infinite
   (d) $\langle \psi | H | \psi \rangle < 0$

155. An electron moving in one dimension ($-\infty < x < \infty$) in a potential of period a is in an eigen state of its Hamiltonian. Its wave function is of the form $u_k(x) \exp(ikx)$. Then $u_k|x'+L\rangle$ is equal to
   (a) $L u_k(x)$
   (b) $u_k(x)$
   (c) $u_k(x-L)$
   (d) $\left(\dfrac{1}{L}\right) u_k(x)$

156. The total wave functions of a system of identical fermions is _____ with respect to interchange of any two particles
   (a) antisymmetric
   (b) symmetric
   (c) Hermitian
   (d) skew symmetric

157. For a spherically symmetric probability cloud of an electron, the _____ quantum number is zero
   (a) principal
   (b) orbital
   (c) spin
   (d) magnetic orbital

158. The wave function of a particle in a classically forbidden region is a
   (a) sine function
   (b) cosine function
   (c) positive exponential
   (d) negative exponential

159. A particle is moving in a Coulomb potential. An operator A commutes with Hamiltonian of the system. The observable corresponding to A is
   (a) position
   (b) linear momentum
   (c) kinetic energy
   (d) angular momentum

160. The possible values of total angular momentum J resulting from the addition of two angular momenta $J_1 = 1$ and $J_2 = 2$ are
   (a) 1, 2
   (b) 1, 3
   (c) 0, 1, 2
   (d) 1, 2, 3

161. The quantization condition for the electron wave is that
   (a) the value of wave function ψ must not be discontinuous
   (b) the value of $\frac{\partial \psi}{\partial x}$ must not be discontinuous
   (c) the value of ψ and $\frac{\partial \psi}{\partial x}$ must be discontinuous
   (d) the value of $\frac{\partial \psi}{\partial x}$ and $\frac{\partial^2 \psi}{\partial x^2}$ must be continuous

162. The eigen value of Hermitian operators are
   (a) imaginary
   (b) indeterminate
   (c) real
   (d) zero

163. If an operator commutes with a Hamiltonian, then the operator is
   (a) constant of motion
   (b) dependent on time
   (c) partially time-dependent
   (d) none of the above

164. The parity of wave function $\psi(x) = e^{-xa^2}$ is
   (a) odd
   (b) even
   (c) partly odd and partly even
   (d) nil

165. The zero point energy of harmonic oscillator is a consequence of
   (a) uncertainty principle
   (b) correspondence principle
   (c) Hamilton's principle
   (d) none of the above

166. The wave function of a particle in a potential box $V = \infty$ for $x < 0$ and $x > a$, $v = 0$ for $0 < x < a$ is
   (a) $A \sin Kx + B \cos Kx$
   (b) $A \cos\left(\frac{n\pi}{a}\right) x$
   (c) $A \sin\left(\frac{n\pi}{a}\right) x$
   (d) $A e^{-Kx}$

167. The difference between the scattering and bound state in quantum mechanical problems is
 (a) the energy $E > 0$ for scattering and $E < 0$ for bound state
 (b) the energy $E = 0$ for scattering and $E > 0$ for bound state
 (c) the energy $E < 0$ for scattering and $E > 0$ for bound state
 (d) the energy $E > 0$ for scattering and $E = 0$ for bound state

168. In Dirac's theory, a constant of motion is the operator
 (a) L_z (b) $L_z + \dfrac{\hbar}{2\sigma_z}$ (c) σ_z (d) $L_z - \dfrac{\hbar}{2\sigma_z}$

169. In Dirac's theory, it is postulated that positions are
 (a) positive energy states
 (b) negative energy states
 (c) holes left in the negative energy status
 (d) none of the above

170. The relativistic Klein–Gordan equation is a valid relativistic equation for
 (a) charged particles having intrinsic angular momentum
 (b) charged particles having no intrinsic angular momentum
 (c) uncharged particles having intrinsic angular momentum
 (d) uncharged particles having no intrinsic angular momentum

171. The eigen operator $\dfrac{\partial^2}{\partial x^2}$ has eigen value -4 corresponding to an eigen function $\sin \alpha x$. The value of x is
 (a) 4   (b) 2   (c) $2i$ (d) $-2i$

172. The commutator $\left[x^2, \dfrac{d}{dx} \right]$ is
 (a) $-2x$ (b) $\dfrac{d^2}{dx^2}$ (c) $x-2$ (d) $2x$

173. For one-dimensional harmonic oscillator, the Hamiltonian is
 (a) $\dfrac{p^2}{m} - \dfrac{kq^2}{2}$ (b) $\dfrac{2m}{p^2} + \dfrac{kq^2}{2}$
 (c) $\dfrac{2m}{p^2} + \dfrac{2}{kq^2}$ (d) $\dfrac{p^2}{2m} + \dfrac{kq^2}{2}$

174. The KE associated with a plane wave is given by

(a) hk

(b) $\frac{1}{2}mk^2$

(c) $\dfrac{h^2 k^2}{8\pi^2 m}$

(d) $\dfrac{h^2 k^2}{8\pi^2 mc^2}$

175. By wave mechanics, every moving particle is associated with a wave. Why cannot one measure or see the wave due to a cricket ball?
    (a) it does not produce wave
    (b) the wavelengths are too small to observe
    (c) the velocity is less
    (d) it is red in colour

176. Given that A and B are Hermitian operators which do not commute
    (a) $(A + iB)$ is Hermitian
    (b) $(A - iB)$ is Hermitian
    (c) AB is Hermitian
    (d) $(A + B)$ is Hermitian

177. Let ψ be an eigen state of the operator S^2 where S is the spin operator of a particle with spin quantum number S.
    (a) ψ cannot be an eigen state of S_z.
    (b) the eigen values of S_z are integers for all S.
    (c) the eigen values of S_z in the state ψ are $+S$ and $-S$.
    (d) the largest possible eigen value of S_z is S.

178. Let ψ be an eigen state of operator A with eigen value λ.
    (a) ψ is an eigen state of A^2 with eigen value λ^2.
    (b) ψ is not necessarily an eigen state of A^2.
    (c) A and A^2 cannot have any common eigen state.
    (d) All eigen states of A^2 must also be eigen states of A.

179. Let $\sigma_x, \sigma_y, \sigma_z$ denote the three Pauli's matrices.
    (a) Only σ_z can be diagonalized.
    (b) $\sigma_x \sigma_y = \sigma_z$.
    (c) $Tr(\sigma_x + \sigma_y) = 0$.
    (d) $\sigma_x + \sigma_y$ commutes with σ_z although σ_x and σ_y do not.

180. Consider the addition of two angular momenta J_1 and J_2.
    (a) The resultant state must have total angular momentum (J_1+J_2).
    (b) The resultant state can have total angular momentum $(J_1 - J_2)$ if $J_1 > J_2$, and $(J_2 - J_1)$ if $J_2 > J_1$.
    (c) If J_1 and J_2 are half odd integers the resultant state can never have total angular momentum zero.
    (d) If J_1 is integral and J_2 is half odd integral, the two angular momenta cannot be added.

181. Let L_x, L_y, L_z denote the orbital angular momentum operators. Then,
    (a) any two of them can be diagonalized simultaneously
    (b) $L_x + iL_y$ is a Hermitian operator
    (c) L_z commute with $L_x^2 + L_y^2$
    (d) $L_x \hat{i} + L_y \hat{j} + L_z \hat{k}$ transforms like an axial vector (pseudo-vector) under the parity transformation $\vec{r} \to -\vec{r}$

182. Let $A\psi = \lambda\psi$ for an operator A, λ being the eigen value, let U denote a unitary operator such that $[U, A] \neq 0$, then
    (a) $U\psi$ is an eigen state of UAU^+ with eigen value λ
    (b) $U\psi$ is an eigen state of UA with eigen value λ
    (c) ψ is an eigen state of UA with eigen value λ
    (d) ψ is an eigen state of U with eigen value λ

183. Given 3 operators A, B and C such that $[A, B] = 0$ and $[B, C] = 0$, we may conclude that
    (a) $[C, A] = 0$
    (b) $[A, BC] = 0$
    (c) $[AB, C^2] = 0$
    (d) $[A^2, B] + [B^2, C^2] = 0$

184. Let $\psi(x, t)$ be the normalized wave function of a particle in a one-dimensional potential. Then,
    (a) $\psi^*(x, t) \cdot \psi(x, t) = 1$.
    (b) $\psi(x, t)$ is a measurable quantity.
    (c) $-i\hbar \dfrac{\partial \psi(x, t)}{dx}$ is a measurable quantity.
    (d) $\int_{-\infty}^{\infty} \psi^*(x, t)\, \psi(x, t)\, dx = 1$ for all t.

185. The plane wave solution $\psi(x) = Ae^{ikx}$ of the Schrodinger equation for a free particle in a dimension
    (a) is normalizable
    (b) is an eigen state of position operator of the particle
    (c) is an eigen state of momentum operator of the particle
    (d) is minimum uncertainty state

186. In order to have an ion beam under a high voltage (1–10 kV), the pressure of the chamber should be about
    (a) 10^{+1} to 10^{-3} torr
    (b) 10^{-1} to 10^{3} torr
    (c) 10^{1} to 10^{-3} torr
    (d) 10^{-1} to 10^{-3} torr

187. The film thickness can be evaluated from the relation
    (a) $t = \dfrac{\rho A}{\omega}$
    (b) $t = \dfrac{\omega}{\rho A}$
    (c) $t = \dfrac{\omega \rho}{A}$
    (d) $t = \dfrac{\omega A}{\rho}$

188. According to classical theory, conductivity, current density, electric field and electron drift velocity are related by the expression
    (a) $\sigma = \dfrac{nev}{\omega}$
    (b) $\sigma = nevE$
    (c) $\sigma = \dfrac{\sigma_e}{vE}$
    (d) $\dfrac{n}{evE}$

189. The mean free path of electron residing near Fermi level region is a function of
    (a) dimension of the metal film
    (b) temperature
    (c) pressure
    (d) direction

190. The electrical resistance of a discontinuous metallic film having island type of structure increases with the
    (a) lowering of its temperature
    (b) increasing of its temperature
    (c) constant temperature
    (d) constant pressure

191. When the dielectric medium is either air or vacuum, then its permittivity is
    (a) 8.85×10^{-10} Farad/m
    (b) 8.85×10^{-12} Farad/m
    (c) 0.885×10^{-9} Farad/m
    (d) 0.885×10^{-12} Farad/m

192. The loss factor is expressed as
    (a) $\tan\delta = j\omega CR$
    (b) $\tan\delta = \dfrac{j}{\omega CR}$
    (c) $\tan\delta = \dfrac{j\omega}{CR}$
    (d) $\tan\delta = \dfrac{1}{j\omega CR}$

193. The spin eigen functions $\alpha(s)$ and $\beta(s)$ are
    (a) orthogonal
    (b) normal
    (c) orthonormal
    (d) hypothetical

194. In the Helium atom, the product of 1s orbitals with $\left(z = \dfrac{27}{16}\right)$ is used to describe the
    (a) excited state
    (b) eigen function of the He
    (c) spin state
    (d) ground state

195. The effective nuclear charge for 1s electron in the He is
    (a) 0.170
    (b) 1.70
    (c) 17.0
    (d) 170

196. The basic equation in the ellipsometric measurement is
    (a) $\dfrac{r_p}{r_s} = \tan\psi \exp(j_s)$
    (b) $\dfrac{r_p}{r_s} = \tan\psi$
    (c) $\dfrac{r_p}{r_s} = \exp(j_s)$
    (d) $\dfrac{r_p}{r_s} = \tan\psi \exp(-j_s)$

197. At steady state evaporation, the equilibrium vapour pressure (p) is given by
    (a) $p = \dfrac{N}{V}$
    (b) $p = \dfrac{N}{VT}$
    (c) $p = \dfrac{N}{VKT}$
    (d) $p = \dfrac{V}{NKT}$

198. The figure of merit is given by
    (a) $Q = -\tan\delta$
    (b) $\tan^{-1}\delta$
    (c) $Q = \dfrac{1}{\tan\delta}$
    (d) $Q = \dfrac{1}{\tan^{-1}\delta}$

199. A product of two functions, both of which are symmetric or antisymmetric will be
    (a) symmetric
    (b) antisymmetric
    (c) zero
    (d) infinity

200. Harte's SCF method is from a shortcoming that the orbital product wave function is not
    (a) symmetric
    (b) antisymmetric
    (c) zero
    (d) normal

201. The effective nuclear charges of $2s$ electron in C is
    (a) 0.325   (b) 0.0325   (c) 3.25   (d) 32.5

202. According to classical theory, current density, electric field and electron drift velocity are related by the expression
    (a) $J = \dfrac{ne}{v}$ (b) $J = nev$ (c) $J = \dfrac{nv}{e}$ (d) $J = \dfrac{n}{ev}$

203. A film resistance is often expressed in terms of its
    (a) ohmic resistance
    (b) thermal lattice vibrations
    (c) sheet resistance
    (d) imperfections

204. Thin discontinuous films generally exhibit
    (a) zero resistivity
    (b) low resistivity
    (c) high resistivity
    (d) infinitive resistivity

205. The dielectric behaviour of a material at high frequencies is closely related to its
    (a) electrical properties
    (b) magnetic properties
    (c) mechanical properties
    (d) optical properties

206. Identify the equivalent of $\int \psi_\alpha^*(\vec{r}) \psi_\beta(\vec{r}) d^3r$
    (a) $\langle \alpha | \vec{r} \rangle \langle \vec{r} | \beta \rangle$
    (b) $\langle \alpha | \vec{r} | \beta \rangle$
    (c) $\langle \beta | \vec{r} | \alpha \rangle$
    (d) $\langle \beta | \vec{r} \rangle \langle \vec{r} | \alpha \rangle$

207. In Schrodinger picture, if all the matrix elements of a dynamical variable Ωs are constant in time, then such a dynamical variable is said to be a
    (a) time-dependent quantity
    (b) constant of inertia
    (c) constant of transformations
    (d) constant of motion

208. In the Stark effect of first excited state of hydrogen atom, the degeneracy is
   (a) completely removed
   (b) not removed at all
   (c) partially removed
   (d) four-fold

209. WKB approximation gives correct eigen values for harmonic oscillator
   (a) for all states of harmonic oscillator
   (b) small values of the quantum number n
   (c) large values of the quantum number n
   (d) only if it has large perturbation

210. Fermi–Golden rule obtained using time-dependent perturbation theory
   (a) is time-dependent
   (b) is time-independent
   (c) depends on time harmonically
   (d) is partially time dependent

211. The process which is not allowed is
   (a) spontaneous emission
   (b) spontaneous absorption
   (c) induced absorption
   (d) induced emission

212. Quantum mechanical scattering cross section for a perfect rigid sphere of radius a is
   (a) $4\pi a^2$
   (b) πa^2
   (c) a^2
   (d) $2\pi a^2$

213. Partial wave analysis is useful in scattering theory
   (a) for any arbitrary potential
   (b) only if the potential is spherically symmetric
   (c) only for short-range potential
   (d) only for screen Coulomb potential

214. Minimum positive energy of a Dirac free electron is
   (a) zero
   (b) $m_0 c^2$
   (c) $\dfrac{m_0 c^2}{2}$
   (d) $2 m_0 c^2$

215. The Dirac free electron negative energy states are completely filled.
   (a) This statement is wrong.
   (b) There are no negative energy states.
   (c) This statement implies Pauli's exclusion principle.
   (d) The statement does not imply Pauli's exclusion principle.

216. Unitary transformation from one representation to another corresponds to
    (a) a rotation of axes in the Hilbert space
    (b) a generalized rotation of state vector in Hilbert space
    (c) contraction of state vector in Hilbert space
    (d) stretching of state vector in Hilbert space

217. The transformation from the Schrodinger picture to Heisenberg picture
    (a) is orthogonal transformation
    (b) is unitary transformation
    (c) is similarity transformation
    (d) is a coordinate transformation

218. Perturbation method can be applied to
    (a) all problems (system)
    (b) systems with large perturbation
    (c) when perturbation is small and ground state properties are known
    (d) when perturbation is small and ground state properties are not known

219. With Stark effect in the first excited state of hydrogen atom the degeneracy is
    (a) completely removed   (b) two-fold
    (c) four-fold   (d) three-fold

220. The Dipole transition is possible if
    (a) $m' = m + 1, l' = 1^*$ (b) $m' = m, l' = l$
    (c) $m' = m + 1, l' = l \pm 1$ (d) $m' = m - 1, l' = l - 1$

221. In the ionization of hydrogen atom which is placed in harmonically time-varying electric field, the final state wave function is represented by
    (a) $U_k = \exp(i\vec{k} \cdot \vec{r})$
    (b) $U_k = \exp\left(\dfrac{iE_k t}{\hbar}\right)$
    (c) $U_k = \exp\left(\dfrac{iE_k t}{\hbar}\right)$
    (d) $U_k = \pi a_0^3 e^{-\dfrac{r}{a}}$

222. Classical scattering cross section for a perfect rigid structure of radius a is

(a) πa^2 (b) $4\pi a^2$ (c) a^2 (d) $2\pi a^2$

223. The scattering cross section is maximum when the phase shift is

(a) zero (b) 90° (c) 45° (d) 180°

224. Energy gap in the Dirac free electron energy spectrum is

(a) zero (b) $m_0 c^2$ (c) $2m_0 c^2$ (d) $-m_0 c^2$

225. Probability density in Dirac theory for a free electron is

(a) $\psi^* \psi$ (b) $\psi^+ \psi$ (c) $\psi \psi^+$ (d) $\alpha \psi^+ \psi$

ANSWERS

1. (a)	2. (a)	3. (b)	4. (c)	5. (d)
6. (a)	7. (d)	8. (c)	9. (a)	10. (b)
11. (d)	12. (b)	13. (c)	14. (c)	15. (a)
16. (a)	17. (d)	18. (b)	19. (a)	20. (c)
21. (d)	22. (a)	23. (b)	24. (c)	25. (c)
26. (a)	27. (d)	28. (b)	29. (b)	30. (a)
31. (b)	32. (a)	33. (b)	34. (a)	35. (d)
36. (a)	37. (d)	38. (c)	39. (a)	40. (c)
41. (b)	42. (a)	43. (a)	44. (d)	45. (c)
46. (d)	47. (b)	48. (a)	49. (a)	50. (a)
51. (b)	52. (b)	53. (a)	54. (a)	55. (b)
56. (d)	57. (b)	58. (a)	59. (b)	60. (a)
61. (c)	62. (c)	63. (d)	64. (b)	65. (c)
66. (a)	67. (a)	68. (c)	69. (b)	70. (b)
71. (c)	72. (a)	73. (a)	74. (b)	75. (d)
76. (a)	77. (b)	78. (b)	79. (a)	80. (a)
81. (c)	82. (a)	83. (c)	84. (d)	85. (a)
86. (b)	87. (b)	88. (c)	89. (b)	90. (d)

91. (d)	92. (b)	93. (d)	94. (b)	95. (d)
96. (a)	97. (a)	98. (a)	99. (a)	100. (c)
101. (a)	102. (b)	103. (a)	104. (b)	105. (d)
106. (d)	107. (b)	108. (c)	109. (c)	110. (c)
111. (c)	112. (c)	113. (c)	114. (c)	115. (d)
116. (a)	117. (b)	118. (d)	119. (a)	120. (b)
121. (a)	122. (b)	123. (a)	124. (b)	125. (c)
126. (c)	127. (a)	128. (a)	129. (a)	130. (a)
131. (b)	132. (a)	133. (a)	134. (a)	135. (a)
136. (a)	137. (d)	138. (d)	139. (a)	140. (c)
141. (a)	142. (d)	143. (c)	144. (c)	145. (c)
146. (c)	147. (b)	148. (b)	149. (d)	150. (a)
151. (c)	152. (b)	153. (b)	154. (d)	155. (c)
156. (a)	157. (b)	158. (d)	159. (d)	160. (d)
161. (a)	162. (c)	163. (a)	164. (b)	165. (a)
166. (c)	167. (a)	168. (b)	169. (c)	170. (a)
171. (b)	172. (a)	173. (d)	174. (b)	175. (b)
176. (d)	177. (b)	178. (a)	179. (c)	180. (a)
181. (a)	182. (a)	183. (b)	184. (d)	185. (c)
186. (d)	187. (b)	188. (a)	189. (b)	190. (a)
191. (b)	192. (d)	193. (c)	194. (d)	195. (b)
196. (a)	197. (c)	198. (c)	199. (a)	200. (b)
201. (c)	202. (b)	203. (c)	204. (c)	205. (d)
206. (a)	207. (d)	208. (c)	209. (a)	210. (b)
211. (b)	212. (d)	213. (b)	214. (b)	215. (c)
216. (a)	217. (c)	218. (c)	219. (d)	220. (c)
221. (a)	222. (a)	223. (b)	224. (c)	225. (b)

6

ELECTROMAGNETIC THEORY AND ELECTRONICS

FORMULAE

1. The gradient has the formal appearance of a vector ∇, multiplying a scalar T; we get,

$$\nabla T = \left(\hat{i}\frac{\partial}{\partial x} + \hat{j}\frac{\partial}{\partial y} + \hat{k}\frac{\partial}{\partial z}\right)T$$

where, ∇ is a vector operator.

2. The fundamental theorem for gradient is

$$\int_{a}^{b}{}_{\text{line}} (\Delta T) \cdot dl = T(b) - T(a)$$

3. The fundamental theorem for divergence is

$$\int_{\text{Volume}} (\nabla \times V) d\tau = \int_{\text{surface}} V \cdot da$$

4. The fundamental theorem for curl is

$$\int_{\text{surface}} (\nabla \times V) \cdot da = \oint_{\text{line}} V \cdot dl$$

5. Gradient in terms of spherical polar coordinates is

$$\nabla T = \frac{\partial T}{\partial r}\hat{r} + \frac{1}{r}\frac{\partial T}{\partial \theta}\hat{\theta} + \frac{1}{r\sin\theta}\frac{\partial T}{\partial \phi}\hat{\phi}$$

6. Divergence in terms of spherical polar coordinates is

$$\nabla \cdot V = \frac{1}{r^2}\frac{\partial}{\partial r}(r^2 V_r) + \frac{1}{r\sin\theta}\frac{\partial}{\partial \theta}(\sin\theta V_\theta) + \frac{1}{r\sin\theta}\frac{\partial V_\phi}{\partial \phi}$$

7. Curl in terms of spherical polar coordinates is

$$\nabla \times V = \frac{1}{r\sin\theta}\left[\frac{\partial}{\partial \theta}(\sin\theta V_\phi) - \frac{\partial V_\theta}{\partial \phi}\right]\hat{r} + \frac{1}{r}\left[\frac{1}{\sin\theta}\frac{\partial V_r}{\partial \phi} - \frac{\partial(V_\phi r)}{\partial r}\right]\hat{\theta}$$

$$+ \frac{1}{r}\left[\frac{\partial}{\partial r}(rV_\theta) - \frac{\partial V_r}{\partial \theta}\right]\hat{\phi}$$

8. Gradient, divergence and curl in terms of cylindrical coordinates are

$$\nabla T = \frac{\partial T}{\partial r}\hat{r} + \frac{1}{r}\frac{\partial T}{\partial \phi}\hat{\phi} + \frac{\partial T}{\partial z}\hat{z}$$

$$\nabla \cdot V = \frac{1}{r}\frac{\partial}{\partial r}(rV_r) + \frac{1}{r}\frac{\partial V_\phi}{\partial \phi} + \frac{\partial V_z}{\partial z}$$

$$\nabla \times V = \left(\frac{1}{r}\frac{\partial V_z}{\partial \phi} - \frac{\partial V_\phi}{\partial z}\right)\hat{r} + \left(\frac{\partial V_r}{\partial z} - \frac{\partial V_z}{\partial r}\right)\hat{\phi} + \frac{1}{r}\left(\frac{\partial(rV_\phi)}{\partial r} - \frac{\partial V_r}{\partial \phi}\right)\hat{z}$$

9. The Dirac delta function is

$$\nabla \cdot \left(\frac{\hat{r}}{r^2}\right) = 4\pi\delta^3(r)$$

Here, $r = \gamma - \gamma_0$.

10. Coulomb's force is

$$F = \frac{1}{4\pi\varepsilon_0}\frac{qQ}{r^2}\hat{r}$$

Here,

ε_0 = permittivity of free space
q, Q = charges

11. The electric field of the source charges is

$$E(P) = \frac{1}{4\pi\varepsilon_0} \sum_{i=1}^{n} \frac{q_i}{r_i^2} \hat{r}_i \overset{(or)}{\Rightarrow} E(P) = \frac{1}{4\pi\varepsilon_0} \frac{q}{r^2} \hat{r}$$

12. The electric field of a line charge is

$$E(P) = \frac{1}{4\pi\varepsilon_0} \int_{\text{line}} \frac{\hat{r}}{r^2} \lambda \, dl$$

Here, λ is the charge per unit length.

The electric field for a surface charge is

$$E(P) = \frac{1}{4\pi\varepsilon_0} \int_{\text{surface}} \frac{\hat{r}}{r^2} \sigma \, da$$

Here, σ is the charge per unit area.
and for a volume charge is

$$E(P) = \frac{1}{4\pi\varepsilon_0} \int_{\text{volume}} \frac{\hat{r}}{r^2} \rho \, dT$$

Here, ρ is the charge per unit volume.

13. Gauss's law in integral form is

$$\oint_{\text{surface}} E \cdot da = \frac{Q_{enc}}{\varepsilon_0} = \frac{1}{\varepsilon_0} \sigma A$$

Here, Q_{enc} is the total charge enclosed within the surface.

Gauss's law in differential form is

$$\nabla \cdot E = \frac{\rho}{\varepsilon_0} \text{ (In vacuum)}$$

$$\nabla \times E = 0 \text{ (Static case)}$$

14. The electric potential is

$$V = \frac{1}{4\pi\varepsilon_0} \frac{q}{r}$$

15. The Poisson's equation is

$$\nabla^2 V = -\frac{\rho}{\varepsilon_0}$$

where, V is the potential.

16. The Laplace's equation is
$$\nabla^2 V = 0$$

17. The work done in moving a charge is
$$W = QV(P)$$
Here, Q = charge.
$$W = \frac{Qq}{4\pi\varepsilon_0 r}$$

18. The energy of a continuous charge distribution is
$$W = \frac{\varepsilon_0}{2} \int_{\text{all space}} E^2 \, d\tau$$
The potential for a continuous charge distribution is
$$V(P) = \frac{1}{4\pi\varepsilon_0} \int \frac{\rho}{r} \, dT$$

19. The dipole contribution to the potential is
$$V_{\text{dip}}(P) = \frac{p \cdot \hat{r}}{4\pi\varepsilon_0 r^2}$$
Here, P = dipole moment $= \sum_{i=l}^{n} q_i r_i'$

20. The pressure in terms of the field just outside the surface is
$$P = \frac{\varepsilon_0}{2} E^2$$

21. The electric dipole whose potential is
$$V = \frac{1}{4\pi\varepsilon_0} \frac{q \cdot s}{r^2}$$
For quadrupole
$$V = \frac{1}{4\pi\varepsilon_0} \frac{q \cdot s}{r^3}$$
For octopole
$$V = \frac{1}{4\pi\varepsilon_0} \frac{q \cdot s}{r^4}$$

22. The induced dipole moment is
$$p = \alpha E$$
Here,
α = atomic polarizability
E = electric field

23. A dipole p in a uniform field E experiences a torque
$$N = p \times E$$
where, p = dipole moment

24. The formula for the force on a dipole in a non-uniform field is
$$F = (p \cdot \nabla)E$$

25. A dipole moment $p = PdT$ in each volume element dT, so the total potential is
$$V = \frac{1}{4\pi\varepsilon_0} \int_{\text{volume}} \frac{P \cdot \hat{r}}{r^2} dT$$

26. The electric displacement is
$$D = \varepsilon_0 E + P$$
(In matter) $\nabla \cdot D = \rho_f$, $\nabla \times D = 0$
$$\oint_{\text{surface}} D \cdot da = Q_{f_{emc}}$$
Here, $Q_{f_{emc}}$ = total charge enclosed in the volume.

27. The polarization proportional to the field is
$$P = \varepsilon_0 \chi_e E \quad \text{(In linear media)}$$
Here, χ_e = electric susceptibility.
$$D = \varepsilon E, \ H = \frac{B}{\mu}, \ M = \chi_m H \quad \text{(linear media)}$$

28. The polarizability in terms of the dielectric constant is
$$\alpha = \frac{3\varepsilon_0}{N} \frac{(K-1)}{(K+2)}$$

29. The formula for susceptibility is

$$\chi_e = \frac{N\dfrac{\alpha}{\varepsilon_0}}{\left(1 - \dfrac{N\alpha}{3\varepsilon_0}\right)}$$

30. The magnetic force on a charge Q, moving with velocity V in a magnetic field B is

$$F_{mag} = Q(V \times B), \; F_{ele} = QE$$

31. The Lorentz force law is

$$F = Q[E + (V \times B)]$$

32. The cyclotron frequency at which the particle would revolve in the absence of any electric field is

$$\frac{QB}{m} = \omega$$

33. The magnetic force on a segment of current-carrying wire is

$$F_{mag} = \int (\lambda dl)(V \times B) = \int (I \times B) dl$$

The magnetic force on a volume current is

$$F_{mag} = \int (\rho dT)(V \times B) = \int (J \times B) dT$$

Here, $J = \rho v$ = current per unit area perpendicular to flow.

34. The continuity equation is

$$\nabla \cdot J = -\frac{\partial \rho}{\partial t}$$

where, ρ = density.

35. The Biot–Savart law is

$$B(P) = \frac{\mu_0}{4\pi} I \int \frac{dl \times \hat{r}}{r^2}$$

Here,

dl = an element of length along the wire
I = current
$\hat{r}$ = distance between the source to the point P
μ_o = permeability of free space

36. The integral of B around a circular path of radius R centred at the wire, would be

$$\oint B \cdot dl = \oint \left(\frac{\mu_o I}{2\pi R}\right) dl$$

$$= \frac{\mu_o I}{2\pi R} \oint dl$$

$$= \mu_o I$$

$$\nabla \times B = \mu_o J \text{ (Ampere's law)}$$

37. The magnetic field at a distance r from a long straight wire, carrying a current I is

$$B = \frac{\mu_o I}{2\pi r}$$

38. The magnetic field of a very long solenoid, consisting of N closely wound turns per unit length on a cylinder of radius R and carrying a steady current I is

$$B = \begin{cases} \mu_o NI\hat{Z} & \text{inside the solenoid} \\ 0 & \text{outside the solenoid} \end{cases}$$

39. The magnetic field of a toroidal coil is

$$B(r) = \begin{cases} \frac{\mu_o nI}{2\pi r}\hat{\phi}, & \text{for points inside the coil} \\ 0, & \text{for points outside the coil} \end{cases}$$

where, n = the total number of turns.

40. Maxwell's equation $\nabla \cdot B = 0$ (For magnetic field)

$$B = \nabla \times A$$

where, A is the vector potential.

$$\nabla \cdot A = 0$$

$$\nabla^2 A = -\mu_0 J$$

$$A = \frac{\mu_0}{4\pi} \int \frac{J}{r} dT \text{ (source)}$$

For line and surface currents,

$$A = \frac{\mu_0}{4\pi} \int \frac{I}{r} dl, \quad A = \frac{\mu_0}{4\pi} \oint \frac{K}{r} da$$

41. The vector potential is "circumferential", using in a circular "amperian loop" at radius r inside the solenoid is

$$A = \frac{\mu_0 NI}{2} r\hat{\phi} \qquad (r < R)$$

For outside,

$$A = \frac{\mu_0 NI}{2} \frac{R^2}{r} \hat{\phi} \qquad (r < R)$$

42. For dipole

$$A_{dip} = \frac{\mu_0}{4\pi} \frac{m \times \hat{r}}{r^2}$$

Here, m = magnetic dipole moment of the loop.

$$m = \frac{1}{2} I \oint (r' \times dl)$$

43. The torque is $N = m \times B$

44. For an infinitesimal loop of dipole moment in a field B, the force is

$$F = \nabla (m \cdot B)$$

45. Magnetic field $H = \dfrac{1}{\mu_0} B - M$

$$\nabla \times H = J_f$$

$$\oint H \cdot dl = I_{f_{enc}}$$

Here, M = magnetization.

46. The current density is
$$J = \sigma f$$
Here, f is the force exerted per unit charge.

47. The electromotive force is
$$\varepsilon = \oint f \cdot dl = IR$$
Here, f is the electrostatic force.

48. The emf or electromotive force generated in the loop is minus the rate of change of flux through the loop.
$$\oint E \cdot dl = \varepsilon = \frac{-d\phi}{dt} \quad \text{(Faraday's law in integral form)}$$

49. The Faraday's law is given by
$$\nabla \times E = -\frac{\partial B}{\partial t} \quad \text{(In vacuum)}$$
Ampere's law with Maxwell's correction is
$$\nabla \times B = \mu_0 J + \mu_0 \varepsilon_0 \frac{\partial E}{\partial t}$$

50. The Neumann formula is
$$M_{21} = \frac{\mu_0}{4\pi} \oiint \frac{dl_1 \cdot dl_2}{r}$$
one integration around loop 1, the other around loop 2.

51. If the current changes, the emf induced in the loop is given by Faraday's law
$$\varepsilon = -L \frac{dI}{dt}$$

52. Energy in magnetic field is
$$\omega = \frac{1}{2\mu_0} \int_{\text{all space}} B^2 d\tau$$

$$\omega_{ele} = \frac{1}{2}\int(V\rho)d\tau$$

$$= \frac{\varepsilon_0}{2}\int E^2 d\tau$$

$$\omega_{mag} = \frac{1}{2}\int(A\cdot J)d\tau$$

$$= \frac{1}{2\mu_0}\int B^2 d\tau$$

53. The self inductance of the coil is

$$L = \frac{\mu_0 I}{2\pi}\ln\left(\frac{b}{a}\right)$$

54. The displacement current is

$$J_d = \varepsilon_0 \frac{\partial E}{\partial t}$$

55. In terms of free charges and currents, the Maxwell's equations read

  (i) $\nabla\cdot D = \rho_f$ (In matter)

  (ii) $\nabla\cdot B = 0$

  (iii) $\nabla\times E = \dfrac{-\partial B}{\partial t}$

  (iv) $\nabla\times H = J_f + \dfrac{\partial D}{\partial t}$

  Here,
    D = displacement current
    J_f = current density

56. The bound current is

$$J_f = \nabla\times M$$

Here, M = magnetization.
The polarization current is

$$J_p = \frac{\partial p}{\partial t}$$

57. Magnetic charge continuity equation is $\nabla \cdot K = \dfrac{-\partial \eta}{\partial t}$

Here,
η = density of magnetic charge
K = current density

58. The boundary conditions for electrodynamics are

(a) $\varepsilon_1 E_{1_\perp} - \varepsilon_2 E_{2_\perp} = \sigma_f$

(b) $B_{1_\parallel} - B_{2_\parallel} = 0$

(c) $B_{1_\perp} - B_{2_\perp} = 0$

(d) $\dfrac{1}{\mu_1} B_{1_\parallel} - \dfrac{1}{\mu_2} B_{2_\parallel} = K_f \times \hat{n}$ (if there is free charge)

59. E in terms of V and A is

$$E = -\nabla V - \dfrac{\partial A}{\partial t}$$

Here, A = scalar potential.

60. The Coulomb gauge is

$$\nabla \cdot A = 0$$

The Lorentz gauge is

$$\nabla \cdot A = -\mu_0 \varepsilon_0 \dfrac{\partial V}{\partial t}$$

The d'Alembertian is

$$\nabla^2 - \mu_0 \varepsilon_0 \dfrac{\partial^2}{\partial t^2} = \square^2$$

$$\square^2 A = -\mu_0 J$$

$$\square^2 V = \dfrac{-\rho}{\varepsilon_0}$$

61. The differential version of Poynting theorem is

$$\nabla \cdot S = \dfrac{-\partial}{\partial t}(U_M + U_{EB})$$

Here,
   U_M = mechanical energy density
   U_{EB} = energy density of the fields

62. The density momentum in the fields is
$$P_{EB} = \mu_0 \varepsilon_0 S$$
Here, S = Poynting vector (momentum in electromagnetic fields)

63.
$$\nabla \cdot (-\vec{T}) = -\frac{\partial}{\partial t}(p + p_{EB})$$
Here,
   $(-\vec{T})$ = momentum flux density
   p = density of mechanical momentum

64.
$$\nabla^2 E = \mu_0 \varepsilon_0 \frac{\partial^2 E}{\partial t^2}$$

$$\nabla^2 B = \mu_0 \varepsilon_0 \frac{\partial^2 B}{\partial t^2}$$

65. The wave equation in vacuum is
$$\nabla^2 f = \frac{1}{v^2} \frac{\partial^2 f}{\partial t^2} \quad \text{(electromagnetic wave equations)}$$

66. In regions of space where there is no charge or current, Maxwell's equations read

(i) $\nabla \cdot E = 0$

(ii) $\nabla \times E = \frac{-\partial B}{\partial t}$

(iii) $\nabla \cdot B = 0$

(iv) $\nabla \times B = \mu_0 \varepsilon_0 \frac{\partial E}{\partial t}$

67. The refractive index is related to the electric and magnetic properties of the material by the equation
$$n = \frac{\sqrt{\varepsilon \mu}}{\varepsilon_0 \mu_0}$$

68. The electric and magnetic fields in a monochromatic plane wave with propagation vector K and polarization $\hat{n}$ are
$$E(r,t) = E_0 e^{i(K \cdot r - \omega t)} \hat{n}$$

$$B(r,t) = \frac{1}{C} E_0 e^{i(K \cdot r - \omega t)} (\hat{K} \times \hat{n})$$

$$= \frac{1}{C} \hat{K} \times E$$

69. For non-static sources,

$$V(r,t) = \frac{1}{4\pi\varepsilon_0} \int \frac{\rho(r', t_r)}{r} d\tau \quad \text{(electromagnetic radiation)}$$

$$A(r,t) = \frac{\mu_0}{4\pi} \int \frac{J(r', t_r)}{r} d\tau$$

where, $\rho(r', t_r)$ = charge density that prevailed at point r' at the retarded time t_r.

70. For electric dipole radiation

$$V(r, \theta, t) = \frac{-p_0 \omega}{4\pi\varepsilon_0 c} \left(\frac{\cos\theta}{r} \right) \sin\omega\left(t - \frac{r}{c}\right) \quad \text{(for electric dipole radiation)}$$

$$A(r, \theta, t) = \frac{-\mu_0 p_0 \omega}{4\pi r} \sin\omega\left(t - \frac{r}{c}\right) \hat{K}$$

$$E = -\nabla V - \frac{\partial A}{\partial t}$$

$$= -\frac{\mu_0 p_0 \omega^2}{4\pi} \left(\frac{\sin\theta}{r} \right) \cos\omega\left(t - \frac{r}{c}\right) \hat{\theta}$$

$$B = \nabla \times A$$

$$= -\frac{\mu_0 p_0 \omega^2}{4\pi c} \left(\frac{\sin\theta}{r} \right) \cos\omega\left(t - \frac{r}{c}\right) \hat{\phi}$$

71. For magnetic dipole radiation

$$A(r,\theta,t) = -\frac{\mu_0 \omega m_0}{4\pi c} \left(\frac{\sin\theta}{r} \right) \sin\omega\left(t - \frac{r}{c}\right) \hat{\phi}$$

$$E = -\frac{\partial A}{\partial t} = -\frac{\mu_0 m_0 \omega^2}{4\pi c} \left(\frac{\sin\theta}{r} \right) \cos\omega\left(t - \frac{r}{c}\right) \hat{\phi}$$

$$\nabla \times A = B$$

$$= -\frac{\mu_0 m_0 \omega^2}{4\pi c^2}\left(\frac{\sin\theta}{r}\right)\cos\omega\left(t - \frac{r}{c}\right)\hat{\theta}$$

72. The Larmor formula is

$$P = \frac{1}{4\pi\varepsilon_0}\frac{2}{3}\frac{q^2 a^2}{c^3}$$

where, P = power.

73. Lienard–Wiechert potentials for a moving point charge are

$$V(r,t) = \frac{1}{4\pi\varepsilon_0}\frac{q}{r\left(1 - \frac{r \cdot v}{c}\right)}$$

$$A(r,t) = \frac{\mu_0}{4\pi}\frac{qv}{r\left(1 - \frac{\hat{r} \cdot v}{c}\right)}$$

$$= \frac{v}{c^2}V(r,t)$$

74. The electric field of a point charge in motion is

$$E(r,t) = \frac{q}{4\pi\varepsilon_0}\frac{r}{(r \cdot u)^3}\left[u(c^2 - v^2) + r(u \times a)\right]$$

The magnetic field of a point charge in motion is

$$B = \frac{1}{c}(\hat{r} \times E) = \frac{1}{c^2}(v \times E)$$

75. Abraham–Lorentz formula for the radiation reaction force is

$$F_{rad} = \frac{1}{4\pi\varepsilon_0}\frac{2}{3}\frac{q^2}{c^3}a$$

Here, $a = a_0 e^{\frac{t}{\tau}}$.

$$\tau = \frac{1}{4\pi\varepsilon_o} \frac{2}{3} \frac{q^2}{mc^3}$$

76. Einstein's velocity addition rules is

$$V_{AC} = \frac{V_{AB}+V_{BC}}{1+\dfrac{V_{AB}+V_{BC}}{C^2}}$$

77. Time dialation $= \Delta t' = \sqrt{1-\dfrac{v^2}{c^2}}\Delta t$

78. The interval recorded on the train clock $\Delta t'$, is shorter by the factor

$$\gamma = \frac{1}{\sqrt{1-\dfrac{\mu^2}{c^2}}}$$

79. Lorentz contraction is

$$\Delta x' = \frac{1}{\sqrt{1-\dfrac{\mu^2}{c^2}}}\Delta x$$

80. The relativistic momentum and energy is represented by

$$p = \frac{mu}{\sqrt{1-\dfrac{\mu^2}{c^2}}}$$

$$E = \frac{mc^2}{\sqrt{1-\dfrac{\mu^2}{c^2}}}$$

$$E^2 - p^2c^2 = m^2c^4$$

81. The current density 4 vector is

$$J^\mu = (c_p, J_x, J_y, J_z)$$

$$\frac{\partial J^\mu}{\partial x^\mu} = 0$$

$$\frac{\partial F^{\mu v}}{\partial x^v} = \mu_0 J^\mu, \quad \frac{\partial G^{\mu v}}{\partial x^v} = 0$$

82. In terms of $F^{\mu v}$ and the power velocity η^μ, the Minkowski force on a charge q is given by

$$K^\mu = q\eta_v F^{\mu v}$$

The 4 vector potential is

$$A^\mu = \left(\frac{V}{c}, A_x, A_y, A_z\right)$$

$$F^{\mu v} = \left(\frac{\partial A^v}{\partial x_\mu} - \frac{\partial A^\mu}{\partial x_v}\right)$$

$$\Box^2 A^\mu = -\mu_0 J^\mu$$

83. The divergence of V in curvilinear coordinates is

$$\nabla \cdot V = \frac{1}{fgh}\left(\frac{\partial}{\partial u}(ghV_u) + \frac{\partial}{\partial u}(fhV_v) + \frac{\partial}{\partial w}(fgV_w)\right)$$

84. The curl of V in terms of curvilinear coordinates is

$$\nabla \times V = \frac{1}{gh}\left[\frac{\partial}{\partial V}(hV_w) - \frac{\partial}{\partial w}(gV_v)\right]\hat{u} + \frac{1}{fh}\left[\frac{\partial}{\partial w}(fV_u) - \frac{\partial}{\partial u}(hV_w)\right]\hat{V}$$

$$+ \frac{1}{fg}\left[\frac{\partial}{\partial u}(gV_v) - \frac{\partial}{\partial V}(fV_u)\right]\hat{w}$$

85. The total energy stores in electromagnetic field is

$$\omega = \frac{1}{2}\int\left(\varepsilon_0 E^2 + \frac{1}{\mu_0}B^2\right)d\tau$$

86. Rate of canonical momentum is

$$\frac{dp}{dt} = -\nabla U$$

Here,
p = canonical momentum
U = potential energy

87. The Poynting vector is

$$\delta = \frac{1}{\mu_o}(E \times B)$$

88. The energy density of the fields is given by

$$U_{EB} = \frac{1}{2}\left(\varepsilon_o E^2 + \frac{1}{\mu_o}B^2\right) \text{ (electric and magnetic)}$$

89. The integral around a closed path is

$$\oint E \cdot dl = 0$$

$$\therefore \nabla \times E = 0 \quad \text{(static charge distribution)}$$

90. The normal derivative of V is

$$\frac{\partial V}{\partial n} = \nabla V \cdot \hat{n}$$

91. Maxwell's equation $\nabla \times H = J_f$

    Power $= P = I^2 R = VI$ (Joule heating law)

92. The boundary condition for vector E is

$$E_{1_\parallel} - E_{2_\parallel} = 0$$

The boundary condition for vector D is

$$D_{1_\perp} - D_{2_\perp} = \sigma_f$$

For vector B is

$$B_{1_\perp} - B_{2_\perp} = 0$$

For vector H is

$$H_{1_\parallel} - H_{2_\parallel} = K_f \times \hat{n}$$

93. The surface density of the force is

$$F_u = \frac{E \cdot D}{2}$$

94. The generalized Ohm's law in the differential form is

$$J = \sigma(E + E^*)$$

where, E^* = extraneous force field.

95. The law of transformation of fields are expressed by the formulae
$$E'_{\parallel} = E_{\parallel}, B'_{\parallel} = B_{\parallel}$$

$$E'_{\perp} = \frac{E_{\perp} + (v_0 \times B)}{\sqrt{1-\beta^2}}, \quad B'_{\perp} = \frac{B_{\perp} - (v_0 \times E)}{\sqrt{1-\beta^2}}$$

96. Magnetic field in a solenoid $B = \mu_0 n I$

Magnetic field at the centre of the loop $B = \frac{\mu_0}{4\pi} \times \frac{2\pi I}{R}$

97. The laws of transformation of the field E and B for $v_0 \ll c$, is
$$E' = E + (V_0 \times B)$$
$$B' = B - \frac{1}{c^2}[V_0 \times E]$$

98. Electromagnetic field invariant is $E^2 - c^2 B^2 = in V$

99. The circulation of vector H in a stationary field $\oint H dl = I$

100. Inductance of a solenoid is $L = \mu \mu_0 n^2 V$

101. Energy of the magnetic field of a current is
$$W = \frac{LI^2}{2}$$
where, L = inductance.

102. Magnetic field energy density $= w = \frac{B \cdot H}{2}$

103. Maxwell's equation in integral form is
$$\oint E \cdot dl = -\int \dot{B} ds$$
$$\oint D ds = \int \rho dV$$
$$\oint H dl = \int \left(J + \frac{\partial D}{\partial t}\right) dS$$

104. Field E in a parallel plate capacitor and at the surface of a conductor is given by

$$E = \frac{\sigma}{\varepsilon_0 \varepsilon}$$

MULTIPLE CHOICE QUESTIONS

1. Reluctance in a magnetic circuit analogues to
   (a) resistance in a direct circuit
   (b) volume of water in hydraulic circuit
   (c) voltage in an alternating circuit
   (d) inductance in an alternating circuit

2. The permeability of a paramagnetic substance is
   (a) slightly less than that of vacuum
   (b) slightly more than that of vacuum
   (c) much more than that of vacuum

3. Magnetic field is increasing through a copper plate. The eddy currents
   (a) help the field increase
   (b) slow down the increase
   (c) do nothing
   (d) create negative field

4. A current loop has magnetic moment m. The torque N in the magnetic field B is given by
   (a) $N = m \times B$
   (b) $N = m \cdot B$
   (c) $N = 0$
   (d) $N = m \pm B$

5. A bar magnet in earth's field will
   (a) move towards the north pole
   (b) move towards the south pole
   (c) experience a torque
   (d) remain stationary

6. Two parallel wires carry currents i_1 and i_2 going in the same direction. The wires
   (a) attract each other
   (b) repel each other

(c) have no force on each other
(d) build up with higher energy

7. What is the drift velocity of an electron in 1-mm copper wire carrying 10 A current?
   (a) 10^{-5} cm/sec
   (b) 10^{-2} cm/sec
   (c) 10 cm/sec
   (d) 10^5 cm/sec

8. What is the average random speed of the electron in a conductor?
   (a) 10^2 cm/sec
   (b) 10^4 cm/sec
   (c) 10^6 cm/sec
   (d) 10^8 cm/sec

9. Which is the correct boundary condition in magnetostatics at a boundary between two different media?
   (a) the component of B normal to the surface has the same value
   (b) the component of H normal to the surface has the same value
   (c) the component of B parallel to the surface has the same value
   (d) the component of H parallel to the surface has the same value

10. The direction of the magnetic field of a long straight wire carrying current is
    (a) in the direction of the current
    (b) radially outward
    (c) along lines circling the current
    (d) in the opposite direction of the current

11. What is the magnetic field due to a long cable carrying 30000 amperes at distance 1 metre?
    (a) 3×10^{-3} Tesla
    (b) 6×10^{-3} Tesla
    (c) 0.6 Tesla
    (d) 6.0 Tesla

12. A current element idl is located at the origin; the current is in the direction of z-axis. What is the x-component of the field at a point $P(x, y, z)$?
    (a) 0
    (b) $\dfrac{-iydl}{(x^2 + y^2 + z^2)^{\frac{3}{2}}}$
    (c) $\dfrac{ixdl}{(x^2 + y^2 + z^2)^{\frac{3}{2}}}$
    (d) $\dfrac{-iydl}{(x^2 + y^2 + z^2)^{\frac{1}{2}}}$

13. A charge placed in front of a metallic plane
    (a) is repelled by the plane
    (b) does not know the plane is there
    (c) is attracted to the plane by the mirror image of equal and opposite charge
    (d) does not undergo any change

14. A dielectric is placed partly into the parallel plate capacitor which is charged but isolated. It feels a force
    (a) of zero
    (b) pushing it out
    (c) pulling it in
    (d) of oscillation

15. Gauss's law would be invalid if
    (a) there were magnetic monopoles
    (b) the inverse square law were not exactly true
    (c) the velocity of light were not universally constant

16. The electric charge can be held in a position of stable equilibrium
    (a) by a purely electrostatic field
    (b) by a mechanical force
    (c) neither of the above
    (d) by a purely magnetostatic field

17. If p is the polarization vector and E is the field, then in the equation $P = \alpha E$, α in general is
    (a) scalar
    (b) vector
    (c) tensor
    (d) none of the above

18. A radially pulsating charged sphere
    (a) emits electromagnetic radiation
    (b) creates a static magnetic field
    (c) can set a nearby electrified particle into motion
    (d) creates a nuclear field

19. A charge radiates whenever
    (a) it is moving in whatever manner
    (b) it is being accelerated
    (c) it is bound in an atom
    (d) it is a rest charge

20. Radiation emitted by an antenna has angular distribution characteristics of dipole radiation when
    (a) the wavelength is long compared with the antenna
    (b) the wavelength is short compared with the antenna
    (c) the antenna has the appropriate shape
    (d) the wavelength is independent of antenna

21. The frequency of a television transmitter is
    (a) 100 kHz   (b) 1 MHz   (c) 10 MHz   (d) 100 MHz

22. A register in a microprocessor is used to
    (a) store a group of related binary digits
    (b) provide random access data memory
    (c) store a single bit of binary information
    (d) convert analog signal into digital signal

23. A coaxial transmission line has an importance of 50 ohms which changes suddenly to 100 ohms. What is the sign of the pulse that returns from an initial positive pulse?
    (a) none   (b) positive
    (c) negative   (d) imaginary

24. A positive pulse is sent into a transmission line which is short-circuited at the other end. The pulse reflected back
    (a) does not exist (= 0)   (b) is positive
    (c) is negative   (d) is imaginary

25. What is the mechanism of discharge propagation in a self-quenched Geiger counter?
    (a) emission of secondary electron from the cathode by uv-quanta
    (b) ionization of the gas near the anode by uv-quanta
    (c) production of metastable states and subsequent de-excitation.
    (d) emission of positron

26. For low noise charge-sensitive amplifiers, FET input stages are preferred over bipolar transistor because
    (a) they have negligible parallel noise
    (b) they are faster
    (c) they have negligible series noise
    (d) they possess simple circuit

27. Using comparable technology, which ADC type has the lowest value for the conversion time divided by the range, $\frac{T_c}{A}$, with T_c = conversion time and $A = 2n$ with n = number of bits.
    (a) flash ADC
    (b) successive approximation converter
    (c) Wilkinson converter
    (d) None of the above

28. A "derandomizer" is a circuit which consists of
    (a) trigger circuit   (b) FIFO memories
    (c) phase-locked loop   (d) counter

29. A discriminator with a tunnel diode can be built with a threshold as low as
    (a) 1 mV   (b) 10 mV   (c) 100 mV   (d) 0.1 mV

30. Pulse with subnanosecond rise time and a few hundred volts amplitude can be produced using
    (a) avalanche transistor   (b) thyratrons
    (c) mechanical switches   (d) klysteron

31. Inside the programming counter of a microprocessor there is
    (a) the address of the instruction
    (b) the address of the data
    (c) the sentence's number of the program
    (d) the address of the ouput

32. A Schmitt trigger has a dead time
    (a) smaller than the pulse width
    (b) about equal to the pulse width
    (c) larger than the pulse width
    (d) zero

33. What is the direct application in standard NIM electronics of De-Morgan Relation $(A \cup B) = A \cup B$?
    (a) transformation of an *OR* unit into an *AND* unit
    (b) inversion of signals
    (c) realization of an "exclusive *OR*"
    (d) realization of "*NOR*"

34. A digital system can be completely fabricated using
    (a) AND and OR gates only
    (b) All NOR gates or all $NAND$ gates
    (c) Neither of the above
    (d) all logic gates

35. When a capacitor is being discharged
    (a) the energy originally stored in the capacitor can be completely transferred to another capacitor
    (b) the original charge decreases exponentially with time
    (c) an inductor must be used
    (d) the charge decreases linearly with time

36. If L = inductance and R = resistance, what unit does $\frac{L}{R}$ have?
    (a) sec
    (b) sec^{-1}
    (c) amperes
    (d) sec^2

37. Two inductances L_1 and L_2 are placed in parallel far apart. The inductance of both is
    (a) L_1
    (b) $L_1 \times L_2 (L_1 + L_2)$
    (c) $(L_1 + L_2) \times \left(\frac{L_1}{L_2}\right)$
    (d) $\frac{(L_1 + L_2)}{L_1 L_2}$

38. An alternating current generator with a resistance of 10 ohms and no reactance is coupled to a load of 1000 ohms by an ideal transformer. To deliver maximum power to the load, what turns ratio should the transformer have?
    (a) 10
    (b) 100
    (c) 1000
    (d) 10,000

39. An electric circuit made up of a capacitor and an inductor in series can act as an oscillator because
    (a) there is always resistance in the wires
    (b) voltage and current are out of phase with each other
    (c) voltage and current are in phase with each other
    (d) the resistance in the wire becomes negative

40. The force in the x-direction between two coils carrying current i_1 and i_2 in terms of mutual inductance M is given by

(a) $i_1 \times \left(\dfrac{di_2}{dx}\right) \times M$

(b) $i_1 i_2 \times \left(\dfrac{dM}{dx}\right)$

(c) $i_1 i_2 \times \left(\dfrac{d^2 M}{dx^2}\right)$

(d) $i_2 \times \left(\dfrac{di_1}{dx}\right) \times M$

41. In order to obtain the Zener effect, the zener diode has to be
    (a) reverse biased
    (b) forward biased
    (c) connected to resistance
    (d) exposed to solar power

42. A transistor amplifier in a "grounded base" configuration has the following characteristics
    (a) low input impedance
    (b) high current gain
    (c) low output impedance
    (d) low current gain

43. It is possible to measure the impedance of co-axial cable
    (a) with an ohmmeter across the cable
    (b) making use of reflection properties of termination by measuring the attenuation of signals through the cable
    (c) with an ammeter along the cable
    (d) with a voltmeter across the cable

44. The transmission of high frequencies in a co-axial cable is determined by
    (a) the impedance
    (b) $\dfrac{1}{(LC)}$ with L and C the distributed inductance and capacitance dielectric losses and skin-effect

45. The high frequency limit of a transistor is determined by
    (a) the increase of noise figure with frequency
    (b) type of circuit (grounded base/emitter/collector)
    (c) mechanical dimensions of the active zones and the drift velocity of charge carriers
    (d) the decrease of noise figure with frequency

46. The frequency response of a single low-pass filter (R-C circuit) can be compensated (ideally)
    (a) exactly only by an infinite series of RC filters
    (b) exactly only by using LC filters
    (c) exactly by a single high-pass RC filter
    (d) exactly only by using π filters

47. The Hall effect has to do with
    (a) the deflections of equipotential lines in a material carrying a current in a magnetic field
    (b) the rotation of the plane of polarization of light going through a transparent solid
    (c) the space charge in electron flow in a vacuum
    (d) induced charges

48. The spectral lines from an atom in a magnetic field are split in the direction of the field; the higher frequency light is
    (a) unpolarized
    (b) linearly polarized
    (c) circularly polarized
    (d) elliptically polarized

49. To go through the ionosphere, an electromagnetic radiation should have a frequency of at least
    (a) 10 Hz
    (b) 104 Hz
    (c) 107 Hz
    (d) 109 Hz

50. According to Poynting theorem, the flow of electromagnetic energy is
    (a) parallel to the electric field
    (b) parallel to the magnetic field
    (c) perpendicular to both magnetic and electric fields at angle $\pm\frac{\pi}{2}$ to the two fields
    (d) perpendicular to the magnetic field

51. The equation $\nabla^2 \phi = -\frac{\rho}{\varepsilon_0}$ is the statement of
    (a) Gauss law
    (b) Poisson's equation
    (c) Laplace equation
    (d) Ampere's law

52. If a magnetic dipole m is in a non-uniform magnetic field B, the force F experienced by the dipole can be shown to be

$F = \nabla \cdot (m \cdot B) = (B \cdot \nabla)m + (m \cdot \nabla)B + B \times (\nabla \times m) + m \times (\nabla \times B)$. Only one of the four terms on the right is not equal to zero. Which is it?

(a) $(B \cdot \nabla)m$
(b) $(m \cdot \nabla)B$
(c) $m \times (\nabla \times B)$
(d) $B \times (\nabla \times m)$

53. For total reflection to occur at the boundary of two media with refractive indices n_1 and n_2, the critical angle is

(a) $\sin^{-1}\left(\dfrac{n_2}{n_1}\right)$
(b) $\sin^{-1}\left(\dfrac{n_2 - n_1}{n_1}\right)$
(c) $\sin^{-1}\sqrt{n_1 n_2}$
(d) $\sin^{-1}(n_2 + n_1)$

54. Identify the UJT symbol

(a)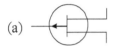
(b)
(c)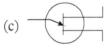
(d)

55. In CRO the time base voltage is of
(a) sawtooth waveform
(b) triangular waveform
(c) rectangular waveform
(d) sinusoidal waveform

56. The use of negative feedback in amplifier results in
(a) higher input resistance
(b) lower output resistance
(c) improvement in frequency response
(d) higher transfer gain

57. What does the following circuit represent?

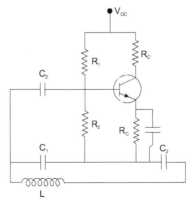

(a) Hartley oscillator (b) Colpitt's oscillator
(c) R-C coupled amplifier (d) astable multivibrator

58. What is the value of the parameter h_{ic} for a transistor in the common-emitter configuration when converted into the common base configuration?

(a) $\dfrac{h_{ob}}{1+h_f b}$ (b) $\dfrac{h_{ib}}{1+h_f b}$ (c) $\dfrac{h_{ob} h_{ib}}{1+h_f b}$ (d) $\dfrac{h_{fb}}{1+h_f b}$

59. The principal property of a varactor diode is
    (a) nearly constant capacitance
    (b) small area, small capacitance
    (c) majority carrier transport
    (d) reactance varies with bias voltage

60. A sinusoidal electromagnetic wave propagating along the positive x-directional plane polarized along the y-direction. The associated E-vector is given by

(a) $B_z = 0, B_y = \dfrac{E_{oy}}{c}\left[\sin\omega t - \dfrac{\omega x}{c}\right]$

(b) $B_y = \dfrac{E_{oy}}{c}\left[\sin\omega t - \dfrac{\omega x}{c}\right]$

(c) $B_y = 0, B_z = \dfrac{E_{oy}}{c}\sin\left[\omega t - \dfrac{\omega x}{c}\right]$

(d) $B_y = 0, B_z = \dfrac{E_{oy}}{c}\sin\left[\omega t - \dfrac{\omega x}{c}\right]$

61. A current amplifier is characterized by
    (a) low input impedance and high output impedance
    (b) high input impedance and low output impedance
    (c) low impedance at both input and output terminals
    (d) high impedance at both input and output terminals

62. Which one of the following characteristics does not belong to a common collector transistor amplifier?
    (a) low voltage gain $[\approx 1]$ (b) high current gain
    (c) high input impedance (d) high output impedance

63. Two identical square frames are made of the same conducting coils such that for ABCD plane is horizontal (with side AB in front) and ADEF plane is vertical, with side AD missing. The same current I flows in the frame as shown in figure. The direction of the magnetic moment vector will be

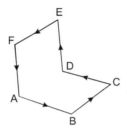

(a) vertical upwards
(b) vertical downwards
(c) at 45° with the vertical pointing upwards
(d) at 45° with the vertical pointing downwards

64. An amplifier has a voltage gain of 500 and an input impedance 20 kΩ, without any feedback. Now a negative feedback with $\beta = 0.15$ is applied. Its gain and input impedance with feedback will respectively be

(a) 9.8 and $392\,\Omega$
(b) 9.8 and 1020 kΩ
(c) 50 and 1020 kΩ
(d) 50 and 2 kΩ

65. The circuit for which the input and output waveforms are shown below is
(a) clipping circuit
(b) integrator
(c) differentiator
(d) Schmitt trigger

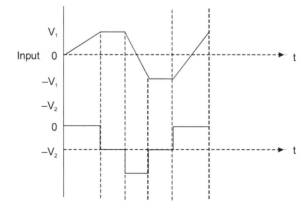

66. Which one of the following situations does not have a resonant charge density $\rho(\eta\, t)$ in space?

(a) $J = 0$

(b) $J = \nabla \times \left(\dfrac{\vec{r}}{r}\right)$

(c) $\vec{J} = \vec{J_0}$ = constant

(d) $\vec{J} = \dfrac{\vec{r}}{r} c;\ c$ = constant, $r \neq 0$

67. Faraday Lenz law relates the rate of change of magnetic flux with emf developed. Which of the following equations represents the above law?

(a) $\nabla \times E = -\dfrac{\partial B}{\partial t}$

(b) $\oint \vec{B} \cdot dl = \varepsilon_0 \mu_0 \dfrac{\partial}{\partial t} \int \vec{E} \cdot \vec{ds}$

(c) $\vec{\nabla} \times \vec{B} = \varepsilon_0 \mu_0 \dfrac{\partial E}{\partial t}$

(d) $\dfrac{\partial p}{\partial t} = q(\vec{E} + \vec{V} \times \vec{B})$

68. The avalanche effect is observed in a diode when
(a) forward voltage exceeds the breakdown voltage
(b) the heavily doped diode is forward-biased
(c) reverse voltage exceeds the breakdown voltage
(d) majority carriers have enough energy to dislodge valence electrons

69. A sphere of radius a has a charge density which varies with distance according to $\rho = Ar^{\frac{1}{2}}$. The electric field at a distance $r < a$ varies with r as

(a) $E \propto r^{-\frac{1}{2}}$ (b) $E \propto r^{\frac{1}{2}}$ (c) $E \propto r^{\frac{3}{2}}$ (d) $E \propto r^{-2}$

70. The vector potential A corresponding to a constant magnetic field B in the z-direction can be represented by

(a) $\dfrac{B}{z}(\hat{i}x - \hat{j}y)$

(b) $\dfrac{B}{z}(\hat{j}x - \hat{i}y)$

(c) $-Bz\hat{k}$

(d) $B(\hat{j}x - \hat{i}y)$

71. In a logic circuit, the output Y in terms of inputs A, B, C is
$$Y = ABC + A\overline{B}C + AB\overline{C}$$
This circuit can be replaced by another logic circuit having
(a) $Y = ABC$
(b) $Y = A(B+C)$
(c) $Y = B(A+C)$
(d) $Y = C(A+B)$

72. The Coulomb force due to two charges is
    (a) directly proportional to their sum
    (b) inversely proportional to their sum
    (c) directly proportional to their product
    (d) inversely proportional to their product

73. The electric field E at any point and the electric flux density D at that point are related by
    (a) $D = \varepsilon E$ (b) $E = \varepsilon D$ (c) $\varepsilon = E \cdot D$ (d) none of the above

74. The force between two current-carrying conductors is
    (a) directly proportional to the distance
    (b) inversely proportional to the distance
    (c) directly proportional to the square of the distance
    (d) inversely proportional to the square of the distance

75. The magnetic field H due to an infinite straight conductor carrying a current I at a perpendicular distance r is
    (a) $\dfrac{I}{r}$ (b) $I \cdot r$ (c) $I \cdot 2\pi r$ (d) $\dfrac{I}{2\pi r}$

76. The capacitance of a parallel plate condenser is
    (a) directly proportional to the area of each plate
    (b) inversely proportional to the area of the plate
    (c) directly proportional to the thickness of each plate
    (d) inversely proportional to the thickness of each plate

77. The Maxwell's equation $\nabla \times E$ could be directly derived from
    (a) Biot–Savart's law
    (b) Ampere's law
    (c) Gauss' law
    (d) Faraday's law

78. The real part of the propagation constant γ is called
    (a) wavelength
    (b) attenuation constant
    (c) phase shift constant
    (d) none of the above

79. Poynting vector is
   (a) $E \cdot H$ (b) $H \cdot E$ (c) $E \times H$ (d) $H \times E$

80. The velocity of electromagnetic waves in air is given by
   (a) $\sqrt{\mu_0 \varepsilon_0}$ (b) $\dfrac{1}{\sqrt{\mu_0 \varepsilon_0}}$ (c) $\sqrt{\dfrac{\mu_0}{\varepsilon_0}}$ (d) $\sqrt{\dfrac{\varepsilon_0}{\mu_0}}$

81. The intrinsic impedance for free space is
   (a) $\sqrt{\mu_0 \varepsilon_0}$ (b) $\dfrac{1}{\sqrt{\mu_0 \varepsilon_0}}$ (c) $\sqrt{\dfrac{\mu_0}{\varepsilon_0}}$ (d) $\sqrt{\dfrac{\varepsilon_0}{\mu_0}}$

82. The power consumed by a TTL IC when compared with corresponding CMOS IC is
   (a) more   (b) less
   (c) equal   (d) more or less depending on particular IC

83. The radii of four-bit Johnson counter is
   (a) 4   (b) 8   (c) 16   (d) 64

84. A parallel-in–parallel-out register shifts
   (a) to the left   (b) to the right
   (c) both to the left and right   (d) does not shift at all

85. The resolution of a 0 to 10 V, 8 bit ADC is
   (a) 10 volts   (b) 1.25 volts
   (c) 39 mV   (d) 19 mV

86. The ADC which uses a DAC within itself is
   (a) flash converter   (b) counter type
   (c) single slope converter   (d) dual slope converter

87. The highest priority interrupt in 8085 is
   (a) TRAP   (b) RST 7.5
   (c) RST 5.5   (d) INTR

88. Push instruction in 8085 sends the data to
   (a) ALU   (b) instruction register
   (c) output device   (d) stacks

89. The gain of ideal op-amp is
   (a) zero   (b) unity   (c) 10^6 (d) infinity

90. The input element used in an exponential amplifier using an op-amp is
    (a) resistor   (b) capacitor
    (c) diode   (d) transistor

91. A scale changer is built using an op-amp with
    (a) resistors only
    (b) resistors and capacitors
    (c) resistors and inductors
    (d) inductors and capacitors

92. Which of the following is universal gate?
    (a) AND   (b) OR   (c) XOR   (d) NOR

93. A radix 4-bit natural up counter counts from 0000
    (a) 1000   (b) 1001   (c) 1010   (d) 1111

94. The time conversion of an 8-bit successive approximation A to D converter using a 1 MHz clock is
    (a) 1 μ sec   (b) 9 μ sec   (c) 8 μ sec   (d) 16 μ sec

95. A 0 to 5 V ADC gives a 12-bit digital O/P. The voltage difference between two successive digital numbers in the O/P approximately corresponds to an analog voltage of
    (a) 10 V   (b) 5 V   (c) 20 mV   (d) 1.2 mV

96. The number of flags available in the 8085 microprocessor is
    (a) indexed   (b) immediate
    (c) indirect   (d) implied

97. The thyristor is a device having alternatively P and N type material and the number of layers is
    (a) 2   (b) 3   (c) 4   (d) 5

98. The direct value of CMRR for good op-amp is
    (a) 0 dB   (b) 10 dB   (c) 20 dB   (d) 120 dB

99. A logarithmic amplifier is built using an op-amp by connecting
    (a) diodes in the input and feedback paths
    (b) diode in the i/p path and a resistance in the feedback path
    (c) resistance in the i/p path and a diode in the feedback path
    (d) resistance in the i/p and feedback paths

100. A circle of radius R has a line charge density $l = l_0 \cos\theta$ where θ is the polar angle (the circle in the xy plane with centre at the origin O). Then

   (a) a potential at O is proportional to $\dfrac{1}{R}$
   (b) the electric field at O is zero
   (c) the potential at O is zero
   (d) the field at O is diverted along $\hat{j}$

101. A spherically symmetric charge distribution is given by
$$\rho(r) = \begin{cases} \dfrac{k}{r} & \text{for } r < a \\ 0 & \text{for } r > a \end{cases} \quad k = \text{constant}$$
Then
   (a) the total charge of distribution is finite
   (b) the total charge of the distribution is infinite
   (c) the dipole moment of the distribution is non-zero
   (d) the total charge of the distribution is indeterminate

102. Consider the 4 Maxwell equations in source-free vacuum
   (a) the four equations can be expressed by a single equation $\partial_\mu F^{\mu r} = 0$, where, $F^{\mu r} = \partial^\mu A^r - \partial^r A^\mu$
   (b) two of the Maxwell equations are represented by $\partial_\mu F^{\mu r} = 0$
   (c) the equation $\partial_\mu F^{\mu r} = 0$ is the Lorentz gauge condition.
   (d) the equation $\partial_\mu F^{\mu r} = 0$ is valid even if sources are present.

103. A charged particle is moving in a region where $\vec{E} = E_0 i$ and $B = B_0 i$. The initial velocity of the particle is $\vec{V} = V_0 k$.

   (a) The force on the particle is directed along $\hat{i}$ at all times.
   (b) The force on the particle is directed along $\hat{j}$ at all times.
   (c) The force on the particle is zero at all times if $V_0 = \dfrac{E_0}{B_0}$.
   (d) The particle will move in the helical path if $V_0 = \dfrac{E_0}{B_0}$.

104. Consider an electromagnetic wave with electric field $\vec{E}$ magnetic field $\vec{B}$ energy density u and Poynting vector $\vec{S}$
    (a) for a monochromatic wave with electric and magnetic contribution to u are equal
    (b) the momentum density of the wave is proportional to $|\vec{S}|^2$
    (c) $\text{div}\vec{S} = 0$
    (d) $u = \dfrac{1}{2}\left[\varepsilon_0 \vec{E}^2 - \dfrac{1}{\mu_0}\vec{B}^2\right]$

105. Consider a rectangular wave guide made of a perfectly conducting material.
    (a) On the inner walls of the wave guide $\vec{E}_{||} = 0$ $\vec{B}_\perp = 0$.
    (b) The electromagnetic waves inside the wave guide are always transverse in nature.
    (c) All waves inside the guide are both transverse electric and transverse magnetic.
    (d) Only TE_{00} modes occur in the wave guide.

106. An infinite long straight wire carrying a steady current I is located along the z-axis (is a closed path of arbitrary shape that encircles the wire once)
    (a) Ampere's law is valid only if c is not planar path (i.e., lies entirely in a plane)
    (b) Ampere's law is valid even if c is not a planar path
    (c) Ampere's law is not valid for steady state
    (d) Ampere's law is valid only if c is circle

107. Consider linear isotropic dielectric media
    (a) if the polarization P is constant everywhere, bound space charge appears
    (b) at the interface between two different dielectrics, no surface charge appears when an external electric field is applied
    (c) Gauss's law for dielectric states that $\oint_s \vec{D} \cdot \vec{ds} = Q$ bound over a closed surface
    (d) if no force charge exists between two dielectrics, electric field lines deviate from the normal when entering the dielectric with a higher permittivity

108. A long thin wire along the z-axis is carrying a steady current I along the positive z-direction. The magnetic field at the point $p = (x, 0, z)$ is

(a) $\dfrac{\mu_0 I}{2\pi x} \hat{e}_y$

(b) $\dfrac{\mu_0 I}{2\pi x} \hat{e}_x$

(c) $\dfrac{\mu_0 I}{2\pi \sqrt{x^2 + z^2}} \hat{e}_y$

(d) $\dfrac{\mu_0 I}{2\pi \sqrt{x^2 + z^2}} \hat{e}_x$

109. A square plane loop of side h is entering a region of constant magnetic field $\vec{B}$ which is perpendicular to the plane of the loop. The loop has its sides parallel to the x-axis and y-axis. The field is in the z-direction. The loop moves in the x-direction with a velocity $V(f)$. The induced emf is

(a) $|B|L|v|$
(b) zero
(c) $\pi |B|L|v|$
(d) infinite

110. Which of the following addressing modes is not available in the 8085 microprocessor
(a) indexed
(b) immediate
(c) indirect
(d) implied

111. According to Gauss's law a conductor which does carry current cannot have an electric field
(a) on its surface
(b) in its interiors
(c) anywhere in its vicinity
(d) anywhere in the space in which it is placed

112. Laplace's equation is applicable for a space
(a) with no charge
(b) with only point charges
(c) with only line charges
(d) with only a volume charge density

113. The capacity of a parallel plate condenser is
(a) directly proportional to the thickness of each other
(b) inversely proportional to the thickness of each other
(c) directly proportional to the distance between the plates
(d) inversely proportional to the distance between the plates

114. The flux density along the axis of a solenoid coil of radius r and length l having N turns and carrying a current I is

(a) $\dfrac{\mu_0 NI}{l}$ (b) $\dfrac{\mu_0 NI}{2\pi r}$ (c) $\dfrac{\mu_0 NI}{2\pi rl}$ (d) $\dfrac{\mu_0 NI}{\pi r^2 l}$

115. A magnetizing fluid H_1 in a medium permeability μ_1 passes normally into a medium with permeability μ_2 as H_2. The relation is

(a) $H_1 = H_2$ (b) $\mu_1 H_1 = \mu_2 H_2$
(c) $\mu_2 H_1 = \mu_1 H_2$ (d) none of the above

116. The energy stores in a coil of inductance L carrying a current I is

(a) $\dfrac{1}{2LI^2}$ (b) $\dfrac{1}{2L^2 I}$
(c) LI^2 (d) $L^2 I$

117. The Maxwell's equation containing $\nabla \times E$ is derived from

(a) Ampere's law (b) Faraday's law
(c) Coulomb's law (d) Gauss' law

118. The velocity of electromagnetic waves in empty space is given by

(a) $\sqrt{\mu_0 \varepsilon_0}$ (b) $\dfrac{1}{\sqrt{\mu_0 \varepsilon_0}}$ (c) $\sqrt{\dfrac{\mu_0}{\varepsilon_0}}$ (d) $\sqrt{\dfrac{\varepsilon_0}{\mu_0}}$

119. The dominant mode in a rectangular waveguide is

(a) TE_{10} (b) TE_{11} (c) TE_{01} (d) TM_{11}

120. The unit for the attenuation constant is
(a) decibels (b) metre/decibels
(c) decibels/metre (d) none

121. The output of a two input EX-OR gate is TRUE, when
(a) both inputs are true (b) either inputs is true
(c) one input is false (d) both inputs are false

122. Simplify the Boolean expression $(a+b)(a+b')$
(a) $a+b$ (b) $ab+c$ (c) a (d) $a+bc$

123. A 5.25" double density floppy disk drive base
(a) 48 tracks/inch (b) 96 tracks/inch
(c) 24 tracks/inch (d) 12 tracks/inch

124. Which is not a peripheral device?
    (a) printer
    (b) floppy drive
    (c) CPU
    (d) plotter

125. The Fan-out of TTL logic family is
    (a) 10   (b) 25   (c) 16   (d) 32

126. Example of a volatile memory is
    (a) IC memory
    (b) magnetic tape
    (c) magnetic core
    (d) ROM

127. A magnetic core has a memory capacity of 8K words of 24 bits each. The number of flip-flop required for MBR is
    (a) 15   (b) 25   (c) 24   (d) 48

128. Addressing mode of an instruction is used
    (a) for stack operations
    (b) to interpret the address field of instruction
    (c) to interpret the operation to be performed
    (d) to execute the instruction

129. The PSW
    (a) is a collection of status bit conditions in the CPU
    (b) is used to interpret the CPU
    (c) gives address to the next instruction to be executed
    (d) gives the return address

130. A peripheral is
    (a) an I/O device
    (b) a part of the main memory
    (c) an I/O processor
    (d) a part of the timer

131. Which of the following gates is not available as an IC?
    (a) Inverter
    (b) XOR
    (c) XNOR
    (d) NOR

132. A counter which counts the sequence 1000, 0100, 0010, 0001, 1000 is called
    (a) Down counter
    (b) up counter
    (c) Johnson counter
    (d) up-down counter

133. The minimum number of gates required to build a half adder is
    (a) 1   (b) 2   (c) 3   (d) 4

134. The fastest A to D converter is
    (a) simultaneous converter   (b) counter type converter
    (c) single-slope converter   (d) dual-slope converter

135. The ratio of the highest to lowest resistance values in a resistive ladder digital to analog converter converting 8 bits input is
    (a) 2   (b) 8   (c) 128   (d) 256

136. The length of the instruction register in 8085 is
    (a) 8 bit   (b) 16 bit   (c) 24 bit   (d) none of the above

137. The instruction which resets the carry flag in 8085 microprocessor is
    (a) STC   (b) CMC   (c) ADD A   (d) ANA A

138. The advantage of LCD over LED is
    (a) high persistance   (b) low power consumption
    (c) fast operation   (d) none of the above

139. The main disadvantage in using an op-amp is
    (a) its low gain   (b) its drift
    (c) its input impedance   (d) its offset voltage

140. An op-amp filter circuit uses
    (a) resistors and capacitors but not inductors
    (b) inductors and capacitors but not resistors
    (c) resistors, capacitors and inductors
    (d) only resistors but not inductors or capacitors

141. The circuit shown in figure acts as a

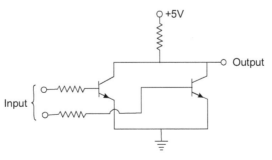

(a) NOR gate (b) NAND gate
(c) AND gate (d) XOR gate

142. A bipolar junction transistor used in common collector configuration has
 (a) high input impedance and high output impedance
 (b) low input impedance and high output impedance
 (c) high input impedance and low output impedance
 (c) low input impedance and low output impedance

143. A sinusoidal voltage $V_0 \sin \omega t$ is applied across a series connection of resistor R and inductor L. The amplitude of the current in this circuit is

(a) $\dfrac{V_0}{\sqrt{R^2 + L^2\omega^2}}$ (b) $\dfrac{V_0}{\sqrt{R^2 - L^2\omega^2}}$

(c) $\dfrac{V_0}{R + L\omega}$ (d) $\dfrac{V_0}{R}$

144. From dimensional argument, the Fourier transform of Coulomb potential in three dimension is

(a) $\dfrac{c}{k^3}$ (b) $\dfrac{c}{k}$ (c) c (d) $\dfrac{c}{k^2}$

where,
 c = constant
 k = magnitude of wave vector

145. The magnetic moment of the current loop carrying current I in an area s is

(a) $\dfrac{Is}{c}$ (b) $\dfrac{Is^2}{c}$ (c) $\dfrac{I^2s}{c}$ (d) $\dfrac{I}{sc}$

146. What are the dimensions of the Poynting vector?
 (a) Energy
 (b) Energy (Area × Time)
 (c) $\dfrac{\text{Energy}}{\text{Area}}$
 (d) $\dfrac{\text{Energy}}{\text{Time}}$

147. An electromagnetic wave incident from a rarer to a denser medium undergoes a phase change of

(a) 0 (b) $\dfrac{\pi}{2}$ (c) π (d) 2π

148. A digital voltmeter should have
    (a) high input impedance (b) low input impedance
    (c) low output impedance (d) high output impedance

149. Which of the following circuits is known as half adder?
    (a) AND (b) OR (c) $NAND$ (d) Exclusive OR

150. The input impedance of an operational amplifier is
    (a) very small   (b) zero
    (c) very high but finite (d) infinite

151. A class B push pull amplifier differs from
    (a) crossover distortion
    (b) harmonic distortion
    (c) intermodulation distortion
    (d) phase delay distortion

152. The operation of JFET involves
    (a) flow of minority carriers
    (b) recombination of charges
    (c) negative resistance
    (d) flow of majority carriers

153. Which of the following waveforms is employed for the horizontal sweep of the electron beam in a CRO?
    (a) triangular   (b) square
    (c) sawtooth   (d) sinusoidal

154. The RC coupled amplifier belongs to
    (a) class AB (b) class C (c) class B (d) class A

155. Which of the following devices is unipolar?
    (a) UJT   (b) FET
    (c) zener diode (d) tunnel diode

156. A closed metallic cubic box of side 1 m has been given a charge of 10^{-2} C. How much work is done in carrying a charge 10^{-6} C from the centre of the cube to one of its corners along the body diagonal?
    (a) $\frac{\sqrt{3}}{12} \times 10^{-8}$ J   (b) 0.75×10^{-8} J
    (c) $\frac{\sqrt{3}}{8} \times 10^{-8}$ J   (d) zero

157. The electric potential in a region is given to be
$$V(x, y, z) = \frac{ax^2 + by^2 + cz^2}{r^3}$$
where, $r^2 = x^2 + y^2 + z^2$. The electric field into the region is

(a) $\left[\dfrac{2z}{\lambda^3} - \dfrac{3(ax^2 + by^2 + cz^2)}{r^5}\right] r$

(b) $\dfrac{(a+b+c)}{3} r$

(c) $a\left[\dfrac{2x}{r^3} - \dfrac{3 x^2}{2 r^5}\right] x + b\left[\dfrac{2y}{r^3} - \dfrac{3 y^2}{2 r^5}\right] y + c\left[\dfrac{2z}{r^3} - \dfrac{3 z^2}{2 r^5}\right] z$

(d) $\dfrac{2(ax + by + cz)}{r^2}$

158. A parallel plate capacitor with plate area 1 cm^2 and a gap of 1 mm is filled with a dielectric material having dielectric constant 3.0. If the plates are given a charge of 10^{-6} C, what is the electrostatic energy stored in the capacitor?

(a) 5×10^{-2} J (b) 18.9×10^{-2} J
(c) 17.3 J (d) 14.6×10^{-2} J

159. Two electric dipoles A and B having dipole strengths P_1 and P_2 respectively are placed at a distance d apart. If A is constrained to lie along x-axis as shown, for what orientation of B will electrostatic energy be minimized?

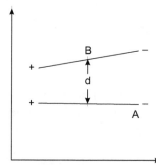

(a) B is parallel to x-axis
(b) B is antiparallel to x-axis

(c) B is along y-axis

(d) B makes an angle $45°$ to x-axis

160. The magnetic field of a current carrying coil of radius b falls with distance r measured along its perpendicular axis for $r >> b$ as

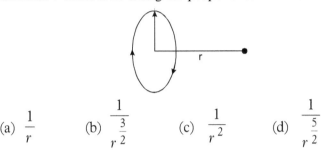

(a) $\dfrac{1}{r}$ (b) $\dfrac{1}{r^{\frac{3}{2}}}$ (c) $\dfrac{1}{r^2}$ (d) $\dfrac{1}{r^{\frac{5}{2}}}$

161. A long solenoid having N turns and radius r_1 contains inside it another solenoid having M turns and radius r_2. If an alternating current I of the frequencies ω is passed through the outer solenoid, the voltage induced in the inner solenoid is proportional to

(a) $\dfrac{MIN}{\omega}$ (b) $\dfrac{MI}{N}\dfrac{\omega r_2}{r_1}$ (c) $\dfrac{MI\omega}{Nr_1}$ (d) $NMI\omega$

162. On a dual trace oscilloscope two waveforms occupy 6 divisions for cycles. Wave B commences 1.2 divisions after commencement of wave A. The sweep setting is 1 m sec/div. What is the frequency and phase difference between A and B?

(a) 333 Hz, 144° (b) 333 Hz, 36°
(c) 167 Hz, 180° (d) 167 Hz, 36°

163. Find the waveforms at the points A and B in the following circuit.

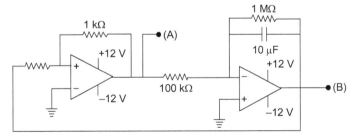

(a) square wave ± 12 V at A and triangular $+ 12$ V at B with periodic time of 1 m sec.

(b) DC $+ 1$ V at A and DC $- 10$ V at B.

(c) triangular + 1 V at A and sawtooth − 10 V at B with periodic time 1m sec.

(d) square wave ±12 V at A and triangular wave ±6 V at B with periodic time 1 m sec.

164. Assuming infinite gain for the op-amp. Calculate the current gain $\dfrac{i_{out}}{i_{in}}$.

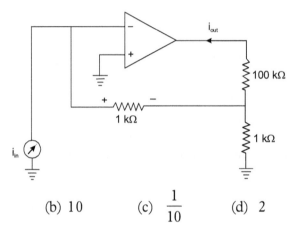

(a) 1   (b) 10   (c) $\dfrac{1}{10}$ (d) 2

165. Find out the voltage V_1 and V_2 in the following p-n-p Si transistor circuit?

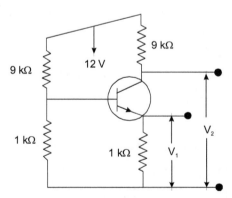

(a) $V_1 = 1.2$ V, $V_2 = 10.8$ V   (b) $V_1 = 0.5V$, $V_2 = 7V$
(c) $V_1 = 1.3$ V, $V_2 = 10.7$ V   (d) $V_1 = 0.7V$, $V_2 = 9V$

166. In the following attenuator circuit a diode is biased by a dc source of 1mA ignoring the small impedances of c compared to 500Ω for input ac signals. Calculate the attenuation (in decibels) for small

input signals. For Si diode use $I = \dfrac{I_0 eV}{nV_t}$ with $n = 2$ and $V_t = 0.025\,V$

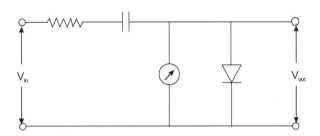

(a) −20 dB (b) −10 dB
(c) −3.8 dB (d) cannot be calculated

167. In photodiodes, photoconductive cells, photovoltaic devices and phototransistors, the principle effect of incident light is
(a) enhanced division of minority carriers towards the inverse biased p–n junction
(b) enhanced recombination of electron-hole pairs reducing device resistance
(c) the excitation of electron across the intrinsic gap of the semi-conductor material
(d) creation of impurity level

168. A static electric field satisfies the equation $\chi E = 0$. The corresponding equation for static magnetic field B is
(a) $B = \mu H$ (b) $B = \chi A$
(c) $\chi B = \mu_0 j$ (d) $B = 0$

169. The addition of gallium to silicon changes
(a) the conductivity but not energy gap width
(b) both conductivity and energy gap width
(c) neither the conductivity nor the energy gap width
(d) the energy gap width but not the conductivity

170. In a phase shift oscillator, the feedback factor must be equal to
(a) unity (b) $\dfrac{1}{29}$
(c) 29 (d) integral multiple of 29

171. The β of the transistor may be determined from the plot of
    (a) V_{CB} vs I_c at constant I_E
    (b) V_{CE} vs I_e at constant I_B
    (c) V_{EC} vs I_E at constant I_B
    (d) V_{BC} vs I_B at constant V_{CE}

172. If an electromagnetic wave is incident on an interface at an angle such that $\tan\theta = n$, the refractive index, then
    (a) both reflected and refracted waves are polarized
    (b) only refracted wave is polarized
    (c) only reflected wave is polarized
    (d) neither wave is polarized

173. The length of the trace (sweep) on the CRT screen is controlled by
    (a) horizontal gain control (b) vertical gain control
    (c) SYNC control (d) trigger level control

174. In a semiconductor diode, the barrier potential offers opposition to
    (a) majority carriers in both regions
    (b) minority carriers in both regions
    (c) free electrons in the n-region
    (d) holes in the p-region

175. A magnetron is
    (a) an instrument to produce microwaves
    (b) an instrument to accelerate charged particles
    (c) a unit of magnetic susceptibility
    (d) a unit of magnetic moment

176. A multi-vibrator circuit which does not require a trigger input is
    (a) bistable multivibrator
    (b) monostable multivibrator
    (c) astable multivibrator
    (d) Schmitt trigger circuit

177. What is the electric field inside a hollow charged sphere of radius a and surface charge density σ at a point at distance r_0 from the centre?
    (a) $\dfrac{a^3\sigma}{3r_0 r^2}$ (b) $\dfrac{r\sigma}{3}$ (c) 0 (d) $\dfrac{r^2\sigma}{3a}$

178. If E and B represent electric and magnetic fields, which of the following statements is correct?
    (a) both E and B are solenoidal
    (b) both E and B are irrotational
    (c) E is solenoidal and B is irrotational
    (d) E is irrotational and B is solenoidal

179. In a conductor the skin depth represents the distance into the conductor where the amplitude E of the wave becomes
    (a) 0
    (b) $\dfrac{E}{c}$
    (c) $\dfrac{E}{\pi}$
    (d) $\dfrac{E}{3}$

180. The power radiating from an oscillating dipole is proportional to
    (a) $\dfrac{p^2}{\lambda^4}$
    (b) $\dfrac{p^4}{\lambda^2}$
    (c) $\dfrac{p^2}{\lambda^2}$
    (d) $\dfrac{p^4}{\lambda^4}$

181. What is the capacitance of a capacitor which holds a charge of 2 Coulomb at a potential of 100 V.
    (a) 0.02 F   (b) 50 µF   (c) 0.02 µF   (d) 200 µF

182. If an ideal op-amp drives no current at both input terminals (1, 2) it means
    (a) $I_1 \neq I_2$
    (b) $I_1 = I_2 = 0$
    (c) $I_1 = -I_2$
    (d) $I_1 = I_2 =$ a small driving current

183. To produce an OR in using NAND gates, the number of gates required is
    (a) 2   (b) 3   (c) 1   (d) 4

184. A full wave rectifier is preferred to half wave rectifier because its
    (a) rectifier efficiency is double and the ripple factor is same as that of the half-wave
    (b) rectifier efficiency and ripple factor are the same as in a half-wave rectifier
    (c) rectifier efficiency is double and ripple factor is low
    (d) rectifier efficiency is low and ripple factor is high

185. The main component responsible for the fall of gain of an RC coupled amplifier in low frequency range is
    (a) the transistor itself
    (b) stray capacitance
    (c) coupling capacitor
    (d) biasing resistors

186. In CRO, the phase difference between two waves can be obtained from
    (a) $\sin\theta = \dfrac{y_1}{y_2}$
    (b) $\sin\theta = \dfrac{y_2}{y_1}$
    (c) $\sin\theta = \dfrac{2y_1}{y_2}$
    (d) $\sin\theta = \dfrac{2y_2}{y_1}$

187. A strain-gauge is a device that converts
    (a) electrical voltage into mechanical displacements
    (b) mechanical displacement into change in resistance
    (c) mechanical displacement into electric current
    (d) electric current into mechanical displacements

188. Among the following types of A/D converters, the most versatile is
    (a) successive approximation type
    (b) counter type
    (c) tracking type
    (d) comparator type

189. Choose the wrong statement concerning the scalar and vector potentials
    (a) for an irrotational field $\vec{F}$, the gradient of the negative of the scalar potential yields $\vec{F}$
    (b) for an irrotational $\vec{F}$, the scalar potential is unique
    (c) for a solenoid field $\vec{F}$ the curl of the vector potential yields $\vec{F}$
    (d) if the field $\vec{F}$ is both irrotational and solenoidal, then the scalar potential does not satisfy the Laplace equation

190. Which of the following statements is not satisfied by the Poynting vector S?
    (a) it represents the energy transported by the electromagnetic fields per unit area per unit time
    (b) $\vec{S}$ lies in the plane containing $\vec{E}$ and $\vec{B}$
    (c) $\vec{S}$ is perpendicular to the plane containing $\vec{E}$ and $\vec{B}$
    (d) electromagnetic fields carry momentum proportional to $\vec{S}$

191. Choose the correct statement concerning the wonderful Maxwell's equations.
    (a) They can be only expressed in differential form.
    (b) They are four in number but can be expressed only in integral form.
    (c) They are not at all consistent with the special theory of relativity.
    (d) They are four in number and can be expressed in both differential and integral forms.

192. Choose the wrong statement concerning the Faraday effect.
    (a) Ionospheric Faraday relation has been measured by satellites, and total electron content of the ionosphere has been determined.
    (b) Faraday rotation of polarized radio waves from pulsar has been observed and has been used to estimate the galactic magnetic field.
    (c) When plane polarized light propagates in the direction of a magnetic field, the plane of polarization will rotate with distance.
    (d) Faraday effect can be observed only in gas and only in the visible range of electromagnetic spectrum.

193. Which of the following statements is not true as far as diffraction is concerned?
    (a) Diffraction emphasizes the wave nature of light.
    (b) A diffraction grating produces sharp maxima in the scattered intensity at all angles.
    (c) Crystals can diffract X-rays, neutrons and slow electrons.
    (d) Structure of extremely complicated biological molecules have been determined by forming their crystals and analysing their diffraction pattern.

194. In a p-n junction diode, the transition capacitance
    (a) decreases with the applied reverse voltage
    (b) increases with the applied reverse voltage
    (c) increases with depletion layer width
    (d) is independent of applied reverse voltage

195. The transformer in amplifiers
    (a) connects the ac output of one stage to the input of next stage while blocking the dc bias voltage
    (b) connects the ac output of one stage to the input of the next stage
    (c) provides the function given in (a) above and also the impedance matching
    (d) provides only the impedance matching

196. In a Schmitt trigger circuit the loop again is
    (a) 0   (b) < 1   (c) 1   (d) > 1

197. The worse case low output of TTL device is (in V)
    (a) 2.4   (b) 0.4   (c) 0.2   (d) 0

198. A multimeter has a resistance of 500 kΩ. This meter is suitable to measure a resistance of the order of (in ohms)
    (a) 5 K   (b) 50 K   (c) 500 K   (d) 5000 K

199. The Coulomb gauge for potential is given by
    (a) $\nabla \times A = B$
    (b) $\nabla \cdot A = 0$
    (c) $\nabla \cdot E = -\dfrac{\partial B}{\partial t}$
    (d) $\nabla \cdot B = 0$

200. The magnetic field at the axis of a pair of Helmholtz coil is most uniform when the separation of the two coils is
    (a) equal to the diameter of the coil
    (b) equal to half of the diameter of the coil
    (c) equal to twice the diameter
    (d) equal to zero

201. When a dielectric sphere is placed in a uniform electric field, the electric field inside the sphere
    (a) becomes 0
    (b) becomes ∞

(c) is parallel to the initial field

(d) is perpendicular to the initial field

202. The electrostatic potential due to a short dipole at large distance is
    (a) inversely proportional to the distance
    (b) directly proportional to the dipole moment
    (c) inversely proportional to the square of the distance
    (d) both (b) and (c) are correct

203. At the boundary between two magnetic media, the boundary condition to be satisfied is
    (a) the normal component of the magnetic induction is continuous
    (b) the tangential component of magnetic induction is continuous
    (c) both the normal and tangential components of magnetic induction are continuous
    (d) the magnetic induction in the medium which has permeability is zero

204. The space charge density at a junction of an unbiased p-n diode is
    (a) zero
    (b) infinite
    (c) negative
    (d) positive

205. If the common emitter transistor is in the saturation region,
    (a) the transistor gets damaged
    (b) the output current becomes independent of input current variation
    (c) the current is constituted mainly by minority carriers
    (d) the efficiency of operation is unity

206. The ratio of fractional change in amplification with feedback to that without feedback is called
    (a) the densitivity
    (b) the feedback ratio
    (c) the sensitivity
    (d) the feedback term

207. In a phase shift oscillator the R-C phase shift network introduces a phase of
    (a) 0°
    (b) 90°
    (c) 360°
    (d) 180°

208. The input offset voltage of an op-amp is the
    (a) small dc voltage to be applied to cause the dc output 0
    (b) maximum voltage that can be applied at the input
    (c) the input voltage for which output is of the same value
    (d) input voltage giving output equal to the supply voltage

209. A digital to analog (D/A) converter is generally located
    (a) at the output of a digital computer
    (b) at the input of the digital computer
    (c) in the CPU (central processor) of a digital computer
    (d) in the memory of a digital computer

210. The circuit below represents

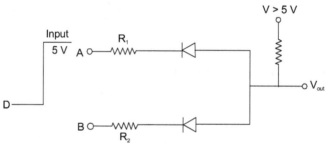

    (a) AND   (b) XOR   (c) OR   (d) NOR
    logic gates for the positive logic.

211. In a constant electric field $\vec{E}$ the electric induction $\vec{D}$ in a certain dielectric medium is found to be given by $\vec{D} = 2.1\varepsilon_0 \vec{E}$. This implies that the dielectric medium is
    (a) linear but not anisotropic
    (b) isotropic but not linear
    (c) linear and isotropic
    (d) non-linear and anisotropic

212. A pulse of electromagnetic radiation propagates through a normally dispersive medium. As it propagates, the pulse gets broadened. This is because
    (a) the medium amplifies the pulse
    (b) the higher frequency components in the pulse travel faster than the lower frequency ones

(c) the lower frequency components in the pulse travel faster than the higher frequency ones

(d) components of all frequencies travel at the same velocity

213. In the logic circuit, the values of A and B for which the output is always equal to 1 (i.e., is always TRUE) are

(a) $A=0, B=0$ (b) $A=0, B=1$

(c) $A=1, B=0$ (d) $A=1, B=1$

214. Storage or diffusion capacitance in a p-n diode arises due to

(a) forward biasing and contribution from minority carriers

(b) forward biasing and contribution from majority carriers

(c) reverse biasing and contribution from majority carriers

(d) reverse biasing and contribution from minority carriers

215. The electric potential in a region of space is given by $\phi = \phi_0 e^{-ax^2}$ where ϕ_0 and a are constant. The charge density in this region is

(a) zero   (b) $2aE_0 x\phi$

(c) $2a\varepsilon_0\phi(1-2ax^2)$ (d) $2a\varepsilon_0\phi(1+2ax^2)$

216. The dispersion relation for EM waves in a certain medium is given by $\omega^2 = ak$ where a is a constant, ω is the frequency and k is the magnitude of the wave vector. Velocity of energy and propagation in this medium is

(a) $\dfrac{2a}{\omega}$ (b) $\dfrac{a}{\omega}$ (c) $\dfrac{a}{2\omega}$ (d) $\dfrac{a}{4\omega}$

217. A point charge q is held at a distance $2a$ from the centre of an isolated uncharged conducting sphere of radius a, the potential of the sphere is

(a) $\dfrac{q}{8\pi\varepsilon_0 a}$ (b) $\dfrac{q}{4\pi\varepsilon_0 a}$ (c) $\dfrac{q}{2\pi\varepsilon_0 a}$ (d) zero

218. The interaction energy of electric dipole p in an external electric $\vec{E}$ is

(a) $\vec{p}\cdot\vec{E}$ (b) $-\vec{p}\cdot\vec{E}$ (c) $|\vec{p}\times\vec{E}|$ (d) $-|\vec{p}\times\vec{E}|$

219. Consider Maxwell's equation in vacuum in the absence of sources. If the solutions of these equations are of the form

$$\vec{E}(\vec{r}_1 t) = \vec{E}_0 \exp[i(\vec{k}\cdot\vec{r} - \omega t)]$$

$$\vec{B}(\vec{r}_1 t) = \vec{B}_0 \exp[i(\vec{k}\cdot\vec{r} - \omega t)]$$

where, $\vec{E}_0, \vec{B}_0$ and k are constant vectors

(a) $\vec{k}\cdot(\vec{E}_0 \times \vec{B}_0) = 0$
(d) $\dfrac{d\omega}{dk} = 0$

(c) $\vec{k}\times(\vec{E}_0 \times \vec{B}_0) = 0$
(d) $\omega = \sigma k$

220. As compared to the single p-n junction diode, a combination of two identical p-n diodes in series
    (a) can withstand twice the value of the peak inverse voltage
    (b) can withstand twice the value of the maximum current
    (c) will have a lower forward resistance
    (d) will have the same forward resistance

221. As the voltage across a forward-biased p-n diode is raised from zero level, the diode resistance
    (a) increases linearly
    (b) decreases linearly
    (c) increases first slowly then exponentially
    (d) decreases first slowly then exponentially

222. Which one of the following properties does not pertain to an operational amplifier
    (a) a high open loop voltage gain
    (b) a high input impedance
    (c) high output impedance
    (d) large CMRR

223. A circular loop of radius a is made of a single turn of thin conducting wire. The self inductance of this loop is L. If the number of turns in the loop is increased from 1 to 8, the self-inductance would be
    (a) $64L$
    (b) $8L$
    (c) $2\sqrt{2}L$
    (d) $\dfrac{L}{8}$

224. A point charge q is held at a distance $\frac{a}{2}$ from the centre of a thin conducting uncharged spherical shell of radius a. The potential of the shell is

(a) $\dfrac{q}{2\pi\varepsilon_0 a}$
(b) $\dfrac{q}{4\pi\varepsilon_0 a}$
(c) $\dfrac{q}{8\pi\varepsilon_0 a}$
(d) zero

225. A coil of n turns kept in a magnetic field experiences zero torque when the place of the coil is
(a) parallel to the field
(b) perpendicular to the field
(c) at 180° to the field
(d) at 270° to the field

226. A changing magnetic field is involved in the operation of
(a) storage batteries
(b) electric irons
(c) flash lights
(d) electric generators

227. Which of the following classes of operation is useful in a linear amplifier?
(a) class A
(b) class C
(c) class D
(d) class E

228. Generally video signals require wider bandwidth than their audio signals, because video signals
(a) are frequency modulated whereas audio signals are amplitude modulated
(b) have a higher information content than audio signals
(c) travel a greater distance than audio signals
(d) are less subject to noise than audio signals

229. The equivalent circuit of a reverse-biased p-n junction circuit has
(a) resistance and capacitance in series
(b) resistance connected in parallel with a capacitor
(c) capacitor only
(d) resistor only

230. The current flowing through the PNP transistor in the figure below is

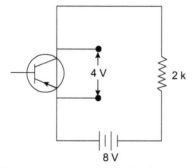

(a) 2 mA   (b) 2.5 mA   (c) 4.5 mA   (d) 6.5 mA

231. Miller effect in a transistor accounts for
   (a) variation in its α
   (b) decrease in its β with frequency
   (c) collector-base feedback
   (d) emitter-base feedback

232. In the figure, midpoint B of the resistor is grounded. The potential at point C is

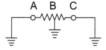

(a) positive
(b) higher than that of B
(c) equal to that of A
(d) negative

233. In a computer, RAM is used as a short memory because it
   (a) is volatile   (b) has small capacity
   (c) is very expensive   (d) is programmable

234. For a high frequency oscillator, a tunnel diode must be biased in its
   (a) first positive-resistance region
   (b) second positive-resistance region
   (c) negative-resistance region
   (d) reverse direction

235. The colour of light emitted by LED depends on
    (a) its forward bias
    (b) its reverse bias
    (c) amount of forward current
    (d) the type of semiconductor material used

236. The main use of class C amplifier is
    (a) as an RF amplifier
    (b) as a stereo amplifier
    (c) in communication sound equipment
    (d) as distortion generator

237. The input gate current of FET is a
    (a) few microamperes
    (b) negligibly small
    (c) a few milliamperes
    (d) a few amperes

238. The expression $\overline{ABC}$ can be simplified to
    (a) $\overline{A} \cdot \overline{B} \cdot \overline{C}$
    (b) $AB + BC + CA$
    (c) $AB + \overline{C}$
    (d) $\overline{A} + \overline{B} + \overline{C}$

239. The majority carriers in n-p-n transistor are
    (a) electrons
    (b) holes
    (c) protons
    (d) neutrons

240. The field $\vec{E}$ is said to be conservative if
    (a) $\vec{\nabla} \cdot \vec{E} = 0$
    (b) $\vec{\nabla} \times \vec{E} = 0$
    (c) $\dfrac{d\vec{E}}{dt} = 0$
    (d) $\dfrac{d^2\vec{E}}{dt^2} = 0$

241. P-N junction offers high resistance when
    (a) forward biased
    (b) reverse biased
    (c) temperature is increased
    (d) used as rectifier

242. The parameters h_{fe} denotes

(a) $\dfrac{\left(\dfrac{\partial I_c}{\partial I_e}\right)}{V_c \text{ constant}}$

(b) $\dfrac{\left(\dfrac{\partial V_c}{\partial I_c}\right)}{V_e \text{ constant}}$

(c) $\dfrac{\left(\dfrac{\partial I_e}{\partial V_c}\right)}{I_c \text{ constant}}$

(d) $\dfrac{\left(\dfrac{I_c}{\partial I_e}\right)}{I_e \text{ constant}}$

243. The common emitter connection is widely used in practical circuits because
   (a) of its high current gain
   (b) of its low current gain
   (c) of its high input and output impedance
   (d) it is independent of α cut-off frequency effects

244. Transistor-resistor logic circuit is preferred over transistor-diode logic circuit in a situation where
   (a) speed of the maximum in and out are not of prime importance
   (b) speed is important
   (c) maximum in and out are important
   (d) noise immunity is important

245. The binary notation of decimal number 20 is
   (a) 01010   (b) 10100   (c) 00100   (d) 01100

246. The Boolean equation of NAND circuit with two inputs is
   (a) $Y = A + B$
   (b) $Y = \overline{AB}$
   (c) $Y = \overline{A}$
   (d) $Y = \overline{A} \cdot B$

247. A sphere is filled uniformly with electric charge. At points within the sphere the field is
   (a) directly proportional to the distance from the centre
   (b) inversely proportional to the distance from the centre
   (c) directly proportional to the charge
   (d) inversely proportional to the charge

248. The force acting on a particle carrying a unit charge and moving in a scalar potential field ϕ is
   (a) $-\text{grad } \phi$ (b) $\text{curl } \phi$ (c) $\text{div } \phi$ (d) $+\nabla \phi$

249. The Lorentz force $\vec{F}$ exerted on a charge q moving with velocity $\vec{v}$ in a magnetic induction B is $q(\vec{v} \times \vec{B})$. The work done by the force F when charge moves a distance x is given by

(a) $2\vec{F}x$ (b) 0 (c) $\dfrac{\vec{F}x}{2}$ (d) $2\int \vec{F}dx$

250. At an uncharged surface of separation between two homogeneous dielectric the normal component of D is
(a) discontinuous
(b) vanishes
(c) indeterminate
(d) continuous

251. The magnetic induction field $\vec{B}$ of a charge q moving with a velocity $\vec{v}$ at a distance $\vec{r}$ from it is given by

(a) $B = \dfrac{\mu_0}{4\pi} q \dfrac{(\vec{V} \times \vec{r})}{|r|^3}$

(b) $B = \dfrac{\mu_0}{4\pi} \dfrac{(\nabla \times \vec{r})q}{|r|^2}$

(c) $\dfrac{\mu_0 \times 4\pi}{q_0(V \times r)}$

(d) $\dfrac{4\pi \, r^2}{\mu_0 \, q}$

252. A simple way of finding the flow of energy in any electromagnetic field is given by
(a) Aries theorem
(b) Laplace's theorem
(c) Stoke's theorem
(d) Poynting theorem

253. The rate of radiation $\left(\dfrac{dE}{dt}\right)$ from an oscillating dipole is

(a) $-\left(\dfrac{4}{3}\right)\left(\dfrac{\pi^2 C}{\lambda^4}\right)\mu_0^2$

(b) $\left(\dfrac{4}{3}\right)\left(\dfrac{\pi C^2}{\lambda^3}\right)\mu_0^2$

(c) $\left(\dfrac{16}{3}\right)\left(\dfrac{\pi C}{\lambda^2}\right)\mu_0^2$

(d) $-\left(\dfrac{16}{3}\right)\left(\dfrac{\pi^4 C}{\lambda^4}\right)\mu_0^2$

254. If the recorded tape is played at half the recording speed, then
(a) pitch of notes will decrease and intensity will also decrease
(b) pitch of notes will decrease and loudness will increase
(c) frequency of notes will increase apparently
(d) frequency of notes will decrease and loudness will also decrease

255. Magnetic field is measured by
   (a) thermopile
   (b) flux meter
   (c) pyrometer
   (d) avometer

256. No current flows between two charged bodies when connected if they have same
   (a) capacity
   (b) potential
   (c) charge
   (d) none of the above

257. A spherical liquid drop has a diameter of 2 cm and is given a charge of 1 μC. The potential of the surface of the drop is
   (a) 9×10^5 volts
   (b) 9×10^9 volts
   (c) 1×10^{-8} volts
   (d) 1×10^{-1} volts

258. A charged particle moves through a magnetic field perpendicular to it. The energy of the particle
   (a) increases
   (b) decreases
   (c) remains unaffected
   (d) becomes zero

259. The graph showing variation of thermo emf with temperature is a
   (a) parabola
   (b) hyperbola
   (c) straight line
   (d) circle

260. A +ve ion of mass M kg and charge e coulombs travels from rest through potential difference of V volts. Its final KE is
   (a) eV Joules
   (b) Joules
   (c) eV
   (d) none of the above

261. A student connects four $\left(\frac{1}{4}\right)\Omega$ cells in series but one cell has terminals reversed. The external resistance is 1Ω if each cell has an emf of 1.5 volts the current flowing is
   (a) 1.0 ampere
   (b) 1.5 amperes
   (c) 2.0 amperes
   (c) 2.5 amperes

262. RC coupling in R-F amplifier suffers from
   (a) Johnson noise
   (b) Bad tuning
   (c) Miller effect
   (d) unnecessary feedback

263. The two rails of a railway track are insulated from each other and the ground is connected to the millivoltmeter. What is the reading of the milli-voltmeter, when a train travels at a speed of

180 km/hour along the track? (Given that horizontal component of earth's magnetic field is 0.2×10^{-4} weber/m² and the rails are separated by 1 m)

(a) 1 mV (b) 0.1 mV (c) 10 mV (d) $\frac{1}{100}$ mV

264. A 6 V battery of negligible internal resistance is used to charge of 2-μF capacitor through a 100-ohm resistor. The time required to obtain 30% of final charge is

(a) 4.6 μs (b) 460 μs (c) 71 μs (d) 0.46 μs

265. A 20-μF capacitor is placed across a generator which has a max emf of 100 V. The reactance when a frequency is 60 Hz is

(a) 1.59 Ω (b) 15.9 Ω (c) 159 Ω (d) 133 Ω

266. For proper functioning of NPN transistor, the voltages V_E, V_B, V_C measured with respect to emitter supply must be

(a) $V_C > V_B > V_E$
(b) $V_B > V_E$ and $V_B > V_C$
(c) $V_E > V_B > V_C$
(d) $V_B > V_C$ and $V_E > V_B$

267. In the case of FET, the transconductance corresponds to

(a) $\frac{1}{I_D}$ (b) $\frac{1}{I_a}$ (c) $\frac{\partial V_D}{\partial V_a}$ (d) $\frac{\partial I_D}{\partial V_g}$

268. The purpose of the crystal oscillator is
(a) to obtain high power oscillations
(b) to obtain pure sine waves
(c) to obtain square wave
(d) to obtain frequency stability

269. Tunnel diode is used
(a) in pulse and digital circuits as a switch
(b) for rectification of AC
(c) as voltage stabilizer
(d) for amplification

270. Wires are used
   (a) to determine high frequency
   (b) as antennas
   (c) to determine wavelength
   (d) in radar

271. The range of radar is 30 km. What should be its pulse repetition frequencies?
   (a) 1000 pulses/sec
   (b) 500 pulses/sec
   (c) 200 pulses/sec
   (d) 10000 pulses/sec

272. The Gauss law for a discrete set of charges is
   (a) $\oint_S \vec{E} \cdot \vec{n} \, ds = \dfrac{q}{\varepsilon_0}$
   (b) $\oint_S \vec{E} \cdot \vec{n} \, ds = 0$
   (c) $\oint_S \vec{E} \cdot \vec{n} \, ds = \left(\dfrac{1}{\varepsilon_0}\right) \sum q_i$
   (d) $\oint_S \vec{E} \cdot \vec{n} \, ds = \sum q_j$

273. The relative permittivity also called dielectric constant is defined as
   (a) $\dfrac{\varepsilon_0}{\varepsilon}$
   (b) $\dfrac{\varepsilon}{\varepsilon_0}$
   (c) $\varepsilon \times \varepsilon_0$
   (d) $(\varepsilon \varepsilon_0)^2$

   where, ε_0 is field in vacuum and ε is field in dielectric.

274. For a charge-free region Poisson's equation reduces to
   (a) $\nabla^2 V = -\dfrac{\rho}{\varepsilon_0}$
   (b) $\nabla^2 V = -\rho$
   (c) $\nabla^2 V = \dfrac{1}{\varepsilon_0}$
   (d) $\nabla^2 V = 0$

275. A dielectric is placed partly into a parallel plate capacitor which is charged but isolated. It feels a force
   (a) of zero
   (b) pushing it out
   (c) pulling it in
   (d) of infinity

276. A charge placed in front of a metallic plate
   (a) is zero
   (b) is repelled by the plane
   (c) does not know the plane is there
   (d) is attracted to the plane by a mirror image of equal and opposite charge

277. What is the magnetic field due to a long cable carrying 30000 amperes at a distance of 1 metre?
    (a) 3×10^{-3} Tesla
    (b) 6×10^{-3} Tesla
    (c) 0.6 Tesla
    (d) 9×10^{-3} Tesla

278. Which appliance uses the largest current?
    (a) TV
    (b) Iron box
    (c) Drill
    (d) Car head lamp bulb

279. Which is the correct boundary condition in magnetostatics at a boundary between two different media?
    (a) the component of H parallel to the surface has the same value
    (b) the component of B normal to the surface has the same value
    (c) the component of H normal to the surface has the same value
    (d) the component of B parallel to the surface has the same value

280. A plane electromagnetic wave of intensity I falls upon a glass plate with index of refraction n. The wave vector is at right angles to the surface (normal incidence). The coefficient of reflection (of the intensity) at normal incidence for a single interface is
    (a) $R = \dfrac{(n-1)^2}{(n+1)^2}$
    (b) $R = \dfrac{(n-1)}{(n+1)}$
    (c) $R = \dfrac{(n+1)^2}{(n-1)^2}$
    (d) $R = \dfrac{(n+1)}{(n-1)}$

281. The number of distinct standing light which can exist between frequencies 1×10^{15} Hz and 1.2×10^{15} Hz in a cavity of volume 1 cm³ is
    (a) 2.26×10^{11}
    (b) 2.26×10^{12}
    (c) 2.26×10^{13}
    (d) 2.26×10^{14}

282. According to Poynting theorem, the flow of EM energy is
    (a) parallel to the electric field
    (b) parallel to the magnetic field
    (c) perpendicular to electric and magnetic field
    (d) perpendicular to magnetic field

283. The equation $\nabla^2 \phi = -\dfrac{\rho}{\varepsilon_0}$ is the statement of
    (a) Gauss law
    (b) Poisson's equation
    (c) Laplace equation
    (d) Ampere's law

284. A TV set is marked 200 W, 250 V. What is the correct fuse rating when the set is connected to a 250 V supply?
    (a) 0.1 A
    (b) 5.0 A
    (c) 1.0 A
    (d) 0.5 A

285. A radiating pulse charge
    (a) emits EM radiation
    (b) creates a static magnetic field
    (c) can set a nearby electrificial particle into motion
    (d) emits acoustic wave

286. Which of the following situations does not have a constant charge density $\rho(r, t)$ in space?
    (a) $\vec{J} = 0$
    (b) $\vec{J} = \vec{J} \times \left(\dfrac{\vec{r}}{r}\right)$
    (c) $\vec{J} + \vec{J_0}$ = constant
    (d) $\vec{J} = \dfrac{\vec{r}}{r} c$

287. A set of lights for a Christmas tree consists of 12 identical 1.5-W lamps connected in series to a 240-V mains supply. What is the voltage across each light?
    (a) 13.2 V
    (b) 20.0 V
    (c) 8.9 V
    (d) 16.0 V

288. The real part of the propagation constant γ is called
    (a) wavelength
    (b) phase shift constant
    (c) attenuation constant
    (d) velocity

289. The resistance of the heating element of an electric fire is 100 Ω. What is the resistance of the heating element of another fire which has twice the power rating for the same voltage?
    (a) 25 Ω
    (b) 100 Ω
    (c) 200 Ω
    (d) 50 Ω

290. Which of the following is an advantage of FTS?
    (a) it is cheaper
    (b) uses broader slits
    (c) Fedfett and Jacquinor advantage
    (d) use of multiple wavelength

291. The vector plot A corresponding to a constant magnetic field B in the z-direction can be represented by

(a) $\dfrac{B}{z}(\hat{i}x - \hat{j}y)$ (b) $\dfrac{B}{z}(\hat{j}x - iy)$

(c) $-Bz\vec{k}$ (d) $B(\hat{j}x - \vec{i}y)$

292. Gauss law would be invalid if
  (a) there were magnetic monopoles
  (b) the inverse square law were not exactly true
  (c) the velocity of light were not universal constant
  (d) $\oint_S \vec{E} \cdot \vec{r}\, ds = 0$

293. If $\vec{P}$ is the polarization vector and $\vec{E}$ is the electric field, then in the equation $\vec{p} = \alpha \vec{E}$, α in general is
  (a) scalar   (b) vector
  (c) tensor   (d) polar

294. In a dielectric medium, the following relation holds good

(a) $\rho' = \left(1 - \dfrac{1}{\varepsilon_r}\right)$ (b) $\rho' = \rho \varepsilon_0$

(c) $\rho' = \rho\left(1 - \dfrac{1}{\varepsilon_r}\right)$ (d) $\rho' = \varepsilon_0 E + P$

295. A perfectly conducting sphere is placed in a uniform electric field pointing in the z-direction. The induced dipole moment of the sphere is
  (a) $P = 4\pi\varepsilon_0 a E_0$ (b) $P = 4\pi\varepsilon_0 a^3 E_0$
  (c) $P = 4\pi\varepsilon_0 a^2 E_0$ (d) $P = 4\pi\varepsilon_0 a^4 E_0$

296. A charge $q = 2\mu C$ is placed at $a = 10$ cm from an infinite grounded conducting plane sheet. The total charge induced on the sheet is
  (a) q (b) $2q$ (c) $-q$ (d) $-2q$

297. The voltage induced between the wing tips of an aircraft due to the vertical component of the earth's magnetic field for $B = 1.5 \times 10^{-5}$ T, wingtip separation = 70 m and the velocity = 1000 km/hr is
  (a) 2.9 V   (b) 0.29 V   (c) 29 V   (d) 0.029 V

298. How many joules of electrical energy does a 1-kW, 240-V electric heater convert to heat energy in one hour?
(a) 1000 J
(b) 60000 J
(c) 240000 J
(d) 360000 J

299. A current loop has magnetic moment m. The torque N in a magnetic field is given by
(a) $N = m \times B$
(b) $N = m \cdot B$
(c) $N = 0$
(d) $N = \text{curl } B$

300. The velocity of light c, ε_0 and μ_0 are related by
(a) $c = \sqrt{\dfrac{\varepsilon_0}{\mu_0}}$
(b) $c = \sqrt{\dfrac{\mu_0}{\varepsilon_0}}$
(c) $c = \sqrt{\dfrac{1}{\varepsilon_0 \mu_0}}$
(d) $c = \dfrac{1}{\varepsilon_0 \mu_0}$

301. The spectral lines from an atom in a magnetic field are split. In the direction of the field, the higher frequency light is
(a) elliptically polarized
(b) circularly polarized
(c) linearly polarized
(d) unpolarized

302. The electric field at any point due to a line charge at a perpendicular distance d is proportional to
(a) d
(b) d^2
(c) $\dfrac{1}{d}$
(d) $\dfrac{1}{d^2}$

303. The electric field inside a conductor carrying no current is
(a) $\dfrac{v}{l}$
(b) $\dfrac{v}{\varepsilon}$
(c) $\dfrac{v}{\varepsilon \cdot l}$
(d) zero

304. The energy stored in a capacitor C with a potential difference V across its plates is
(a) CV
(b) CV^2
(c) $\dfrac{1}{2}CV^2$
(d) $\dfrac{1}{2}CV$

305. The capacitance of a parallel plate condenser separated by a dielectric of thickness t is inversely proportional to
(a) t
(b) $\dfrac{1}{t}$
(c) t^2
(c) $\dfrac{1}{t^2}$

306. According to Faraday's law of induction the voltage induced in an N-turn coil is proportional to
    (a) rate of change of flux
    (b) flux
    (c) rate of change of flux linkages
    (d) flux linkages

307. At a boundary of two magnetic materials of permeability μ_1 and μ_2 the normal components of H bear the relation
    (a) $\mu_1 H_{n_1} = \mu_2 H_{n_2}$
    (b) $\mu_1 H_{n_2} = \mu_2 H_{n_1}$
    (c) $H_{n_1} = H_{n_2}$
    (d) $H_{n_1} = -H_{n_2}$

308. The Maxwell's equation for $\nabla \times H$ is obtained from
    (a) Ampere's law
    (b) Faraday's law
    (c) Gauss law
    (d) Coulomb's law

309. The imaginary part of the propagation constant γ gives
    (a) attenuation constant
    (b) phase constant
    (c) velocity of the plane wave
    (d) wavelength

310. dB_m is the unit used to measure
    (a) wavelength
    (b) velocity of electromagnetic wave
    (c) RF signal strength
    (d) attenuation

311. The first waveguide was proposed by
    (a) J. J. Thomson
    (b) O. L. Lodge
    (c) Lord Rayleigh
    (d) Mc Lachan

312. Two identical point charges separated by a distance have a force of
    (a) attraction
    (b) repulsion
    (c) could be either attraction or repulsion
    (d) have no force at all

313. The ratio of electric field E_1 to the electric flux density at any point is
    (a) zero
    (b) infinity
    (c) ε_1 the permittivity
    (d) $\dfrac{1}{\varepsilon}$

314. The electric field due to a surface charge at a point at a distance d is
    (a) directly proportional to d
    (b) inversely proportional to d
    (c) inversely proportional to d^2
    (d) independent of d

315. A capacitor having a charge Q and a capacitance C has a stored energy
    (a) CQ
    (b) $\dfrac{1}{2}CQ$
    (c) $\dfrac{1}{2}CQ^2$
    (d) $\dfrac{1}{2}\dfrac{Q^2}{C}$

316. If a current I produces a flux ϕ on an N-turn coil, the self inductance of the coil is
    (a) $\dfrac{N\phi}{I}$
    (b) $\dfrac{NI}{\phi}$
    (c) $\dfrac{I}{N\phi}$
    (d) $\dfrac{\phi}{NI}$

317. At a boundary of two magnetic materials of permeability μ_1 and μ_2 the tangential components of H bear the relation
    (a) $\mu_1 H t_1 = \mu_2 H t_2$
    (b) $\mu_1 H t_1 = \mu_2 H t_2$
    (c) $H t_1 = H t_2$
    (d) $H t_1 = -H t_2$

318. The Maxwell's equation for $\nabla \times E$ is derived from
    (a) Ampere's law
    (b) Faraday's law
    (c) Gauss law
    (d) Coulomb's law

319. The real part of the propagation constant γ gives
    (a) attenuation constant
    (b) phase constant
    (c) amplitude of the wave
    (d) frequency

320. Poynting vector is given by
    (a) $E \times H$
    (b) $H \times E$
    (c) $H \cdot E$
    (d) $E \cdot D$

321. The dominant mode in a circular waveguide is
    (a) TM_{10}
    (b) TM_{01}
    (c) TE_{10}
    (d) TE_{11}

322. The logic family which consumes least power is
(a) TTL (b) ECL (c) I²L (d) CMOS

323. The gate which is not available as a CMOS IC is
(a) XOR (b) XNOR (c) NOT (d) NAND

324. Number of flip-flops required to build a radix 25 counter is
(a) 3 (b) 4 (c) 5 (d) 6

325. The fastest ADC is
(a) simultaneous converter (b) dual slope converter
(c) single slope converter (d) counter converter

326. Which of the following addressing modes is not available in 8085?
(a) Immediate (b) Indirect
(c) Implied (d) Indexed

327. Interrupt mastering is done in 8085 using the instruction
(a) RST (b) SIM (c) XCHG (d) PCHL

328. The number of layers on an SCR is
(a) 2 (b) 3 (c) 4 (d) more than 4

329. The input impedance of one ideal op-amp is
(a) 0 (b) finite and low
(c) finite and high (d) infinite

330. The number of integrations needed in an analog computer to solve the 2nd order differential equation is
(a) 2 (b) 3 (c) 4 (d) 5

331. Which of the following is the transfer function of a high pass filter?
(a) $\dfrac{1}{s+1}$ (b) $\dfrac{s}{s+1}$ (c) $\dfrac{1}{s^2+1}$ (d) $\dfrac{s}{s^2+1}$

332. The fastest logic family is
(a) TTL (b) ECL (c) I^2L (d) CMOS

333. Which of the following is a universal gate?
(a) NOT (b) NOR (c) XOR (d) XNOR

334. The number of flip-flops required to build a radii 60 counter is
(a) 3 (b) 4 (c) 5 (d) 6

335. The resolution of an 8-bit −5 to +5 V ADC is about
   (a) 10 volts   (b) 5 volts
   (c) 125 volts   (d) 39 millivolts

336. Which of the following flags is not available in 8085?
   (a) carry   (b) overflow   (c) sign   (d) zero

337. BCD addition in 8085 is possible using
   (a) DAD   (b) DAA   (c) XCHG   (d) AND

338. The number of terminals in a thyristor is
   (a) 2   (b) 3   (c) 4   (d) more than 4

339. The gain of an ideal op-amp is
   (a) 0   (b) 1000   (c) 10^6 (d) infinity

340. The feedback element in an op-amp integrator is a
   (a) resistor   (b) inductor
   (c) capacitor   (d) diode

341. Which of the following is the transfer function of a low pass filter
   (a) $\dfrac{1}{s+1}$ (b) $\dfrac{s}{s+1}$ (c) $\dfrac{1}{s^2+1}$ (d) $\dfrac{s}{s^2+1}$

342. Both the differential and the Thomson scattering cross sections are
   (a) completely independent   (b) completely dependent
   (c) partially dependent   (d) partially independent of the frequency of the incident radiation

343. Thomson scatter cross section is
   (a) $\sigma_T = \dfrac{8\pi}{3} r_e^3$ (b) $\sigma_T = \dfrac{8\pi}{3} r_e$
   (c) $\sigma_T = \dfrac{4\pi}{3} r_e^2$ (d) $\sigma_T = \dfrac{8\pi}{3} r_e^2$

344. The decay time for radiation from an oscillating dipole is
   (a) $\tau = \dfrac{3hc^3}{8\omega^2 \mu_0^2}$ (b) $\tau = \dfrac{3hc^3}{8\omega^2 \mu_0}$
   (c) $\tau = \dfrac{3hc^3}{\omega^2 \mu_0^2}$ (d) $\tau = \dfrac{\hbar c^3}{\omega^2 \mu_0}$

345. Thomson scattering cross section works to
    (a) 2.65×10^{-29} m²
    (b) 6.65×10^{-29} m²
    (c) 5.65×10^{-20} m²
    (d) 10.65×10^{-32} m²

346. According to Coulomb's law the force between two charges q_1 and q_2 is proportional to
    (a) $q_1 + q$
    (b) $q_1 - q_2$
    (c) $q_1 * q_2$
    (d) $\dfrac{q_1}{q_2}$

347. The electric flux density D at any point where the electric field E is proportional to
    (a) E
    (b) E^2
    (c) $\dfrac{1}{E}$
    (d) $\dfrac{1}{E^2}$

348. The electric field inside a conductor which does not carry current is
    (a) infinity
    (b) very small
    (c) small and finite
    (d) zero

349. In a parallel plate condenser if the area of cross section of each plate is doubled, the capacitance value would be
    (a) the same
    (b) doubled
    (c) halved
    (d) four times

350. A solenoid coil of N turns having an axial length l carries a current I. The magnetizing force along its axis is
    (a) NIl
    (b) $\dfrac{Il}{N}$
    (c) $\dfrac{l}{NI}$
    (d) $\dfrac{NI}{l}$

351. Two magnetic materials with permeability μ_1 and μ_2 have common boundary. If the normal component of the field in the first medium is Hn_1 and in the second medium is Hn_2 then
    (a) $Hn_1 = Hn_2$
    (b) $\mu_1 Hn_1 = \mu_2 Hn_2$
    (c) $\mu_2 Hn_1 = \mu_1 Hn_2$
    (d) none of the above

352. The self inductance of the coil with N turns and producing a flux ϕ with current I is
    (a) $N\phi I$
    (b) $\dfrac{NI}{\phi}$
    (c) $\dfrac{N\phi}{I}$
    (d) $\dfrac{I}{N\phi}$

353. Maxwell's equation for $\nabla \times H$ is derived from
     (a) Ampere's law
     (b) Gauss's law
     (c) Biot–Savat's law
     (d) Faraday's law

354. The attenuation constant α of a plane wave is measured in
     (a) metres
     (b) Nepers
     (c) decibels
     (d) seconds

355. If the E field is along the Z-direction and the H field is along the Y-direction, the propagation of the plane wave is along the
     (a) Z-direction
     (b) Y-direction
     (c) X-direction
     (d) 45° to both Y and Z direction

356. According to Coulomb's law, the force between two charges separated by a distance d is proportional to
     (a) d
     (b) d^2
     (c) $\dfrac{1}{d}$
     (d) $\dfrac{1}{d^2}$

357. The electric flux density D at any point and the electric field E at the point are related by
     (a) $D = \varepsilon E$
     (b) $E = \varepsilon D$
     (c) $D^2 = \varepsilon E^2$
     (d) $E^2 = \varepsilon D^2$

358. The electric field at the surface of a conductor which does not carry current is
     (a) zero
     (b) normal to the surface
     (c) tangential to the surface
     (d) inclined to the surface

359. The energy stores in a capacitance C charged to Q coulombs is
     (a) $\dfrac{1}{2}CQ^2$
     (b) CQ^2
     (c) $\dfrac{Q^2}{C}$
     (d) $\dfrac{Q^2}{2C}$

360. A straight conductor carrying a current I produces a field H at a perpendicular distance r. The magnitude of H is
     (a) $\dfrac{I}{r}$
     (b) $\dfrac{I}{2r}$
     (c) $\dfrac{I}{2\pi r}$
     (d) $\dfrac{I}{\pi r}$

361. The voltage induced by an N turn coil when the flux ϕ linking with changes to 0 in time T is
     (a) $\dfrac{\phi}{NT}$
     (b) $\dfrac{N\phi}{T}$
     (c) $\dfrac{NT}{\phi}$
     (d) $\dfrac{T}{N\phi}$

362. The Maxwell's equation for $\nabla \times E$ is derived from
    (a) Ampere's law
    (b) Faraday's law
    (c) Gauss's law
    (d) Coulomb's law

363. If the propagation constant of an electromagnetic wave $\upsilon = \alpha + j\beta$ then α is called
    (a) real propagation constant
    (b) phase constant
    (c) attenuation constant
    (d) none of the above

364. The wavelength λ of the plane wave at frequency f is (c is the velocity of em waves)
    (a) $\dfrac{I}{f}$
    (b) $\dfrac{c}{f}$
    (c) $\dfrac{f}{c}$
    (d) c^*f

365. The number of flip-flops required to build a modulus 100 counter is
    (a) 4
    (b) 6
    (c) 7
    (d) 8

366. The main disadvantages of serial transfer is
    (a) errors are more
    (b) consumes more power
    (c) takes more time
    (d) none of the above

367. The fastest A to D converter is
    (a) dual slope converter
    (b) flash converter
    (c) single slope converter
    (d) successive approximation converter

368. The instruction used for BCD addition in 8085 is
    (a) ADD
    (b) ADC
    (c) DAD
    (d) DAA

369. The device used in a voltage regulator is
    (a) p-n junction diode
    (b) tunnel diode
    (c) Schottky barrier diode
    (d) zener diode

370. When an op-amp works as an inverter, the feedback element is
    (a) resistor
    (b) inductor
    (c) capacitor
    (d) diode

371. The logic family whose speed is best is
    (a) CMOS
    (b) TTL
    (c) I^2L
    (d) ECL

372. The gate which is not available in IC form is
    (a) NAND (b) NOR (c) XOR (d) XNOR

373. Number of flip-flops required to build a modulo 1000 counter is
    (a) 8 (b) 10 (c) 16 (d) 32

374. Registers are builds using
    (a) transistor
    (b) gates
    (c) flip-flops
    (d) capacitors

375. The instruction used in 8085 to mask interrupts is
    (a) PUSH (b) STC (c) DI (d) SIM

376. The number of P and N layers in an SCR is
    (a) 2 (b) 3 (c) 4 (d) more than 4

377. The gain of an ideal op-amp is
    (a) zero (b) unity (c) 10^6 (d) infinity

378. When an op-amp works as an integrator, the feedback element is
    (a) capacitor
    (b) diode
    (c) resistor
    (d) inductor

379. Gauss's law would be invalid if
    (a) these were magnetic monopoles
    (b) inverse square law were not exactly true
    (c) the velocity of light were not universal constant
    (d) $\oint \vec{E} \cdot \vec{n} \, dr = 0$

380. If $\vec{P}$ is the polarization vector and E is the electric field, then in the equation $\vec{p} = \alpha \vec{E}$, α in general is
    (a) scalar
    (b) vector
    (c) tensor
    (d) polar

381. In a dielectric medium, the following relation holds good
    (a) $\rho' = \left(1 - \dfrac{1}{\varepsilon_r}\right)$
    (b) $\rho' = \rho \varepsilon_0$
    (c) $\rho' = \rho\left(1 - \dfrac{1}{\varepsilon_r}\right)$
    (d) $\rho' = \varepsilon_0 E + P$

382. A perfectly conducting sphere is placed in a uniform electric field pointing in the Z-direction. The induced moment of the sphere is
   (a) $P = 4\pi\varepsilon_0 a E_0$
   (b) $P = 4\pi\varepsilon_0 a^3 E_0$
   (c) $P = 4\pi\varepsilon_0 a^2 E_0$
   (d) $P = 4\pi\varepsilon_0 a^4 E_0$

383. A charge q is placed at 10 cm from an infinite grounded conducting plane sheet. The total charge induced on the sheet is
   (a) q
   (b) $2q$
   (c) $-q$
   (d) $-2q$

384. How is an SCR similar to the diode rectifier?
   (a) both can be classified as thyristors
   (b) both support only one direction of current flow
   (c) both are used to change ac to pulsating dc
   (d) they both have one PN junction

385. In a JFET
   (a) both types of majority carriers take part in conduction
   (b) only one type of minority carriers takes part in conduction
   (c) both types of minority carriers take part in conduction
   (d) only one type of majority carriers takes part in conduction

386. For a rectangular wave in the input, an integrating circuit in the output will show
   (a) triangular wave
   (b) square wave
   (c) bistable
   (d) none of the above

387. Due to tunnel diode, a decrease in current causes
   (a) voltage constant
   (b) decrease in voltage
   (c) increase in voltage
   (d) none of the above

388. An ideal op-amp draws
   (a) zero current
   (b) sum of the two input currents
   (c) difference of the two input currents
   (d) infinite current

389. The gain of an inverting op-amp is
   (a) always greater than one
   (b) always less than one
   (c) always equal to one
   (d) greater or less than or equal to one

390. In a uniformly doped p-n junction, the doping level of the p-side is four times the doping level of the n-side. The ratio $\dfrac{w_p}{w_n}$ of the depletion layer width in the p-and n-region is
    (a) 0.25   (b) 0.5   (c) 2   (d) 4

391. In an EMOSFET, the polarity of inversion layer is the same as that of the
    (a) charge on the gate electrode
    (b) minority carries in the drain and source
    (c) majority carries in the substrate
    (d) majority carries in the source and drain

392. The fastest ADC is
    (a) counter-comparator type   (b) successive-approximation type
    (c) parallel-comparator type   (d) dual slope integrating type

393. In a conductor, the skin depth represents the distance into the conductor, where the amplitude E of a wave becomes
    (a) 0   (b) $\dfrac{E}{e}$ (c) E (d) $\dfrac{E}{3}$

394. The dimensions of the Poynting vector are
    (a) energy   (b) energy/time
    (c) energy/area   (d) energy/(area*time)

395. The four types of Bravais lattice—simple, base centred, body centred and face centred—exist is only one crystal system. Identify the system
    (a) trigonal   (b) tetragonal
    (c) orthorhombic   (d) cubic

396. When a dielectric sphere is placed in a uniform electric field, the electric field inside the sphere
    (a) becomes 0
    (b) becomes > 0
    (c) is parallel to the initial field
    (d) is perpendicular to the initial field

397. The magnetic moment of a current loop carrying current I in an area S is
    (a) $\dfrac{IS}{C}$ (b) $\dfrac{IS^2}{C}$ (c) $\dfrac{I^2S}{C}$ (d) $\dfrac{I}{SC}$

ANSWERS

1. (a)	2. (b)	3. (b)	4. (a)	5. (c)
6. (a)	7. (b)	8. (c)	9. (a)	10. (c)
11. (b)	12. (b)	13. (c)	14. (c)	15. (b)
16. (c)	17. (c)	18. (a)	19. (b)	20. (a)
21. (d)	22. (a)	23. (b)	24. (c)	25. (c)
26. (a)	27. (a)	28. (c)	29. (c)	30. (a)
31. (a)	32. (b)	33. (c)	34. (b)	35. (b)
36. (a)	37. (b)	38. (a)	39. (b)	40. (b)
41. (a)	42. (a)	43. (b)	44. (b)	45. (c)
46. (c)	47. (a)	48. (c)	49. (c)	50. (c)
51. (b)	52. (b)	53. (a)	54. (b)	55. (a)
56. (d)	57. (b)	58. (b)	59. (d)	60. (d)
61. (a)	62. (d)	63. (c)	64. (b)	65. (c)
66. (b)	67. (a)	68. (c)	69. (c)	70. (a)
71. (b)	72. (c)	73. (a)	74. (d)	75. (a)
76. (a)	77. (d)	78. (b)	79. (c)	80. (b)
81. (c)	82. (a)	83. (c)	84. (d)	85. (c)
86. (b)	87. (a)	88. (d)	89. (d)	90. (c)
91. (b)	92. (d)	93. (d)	94. (c)	95. (b)
96. (a)	97. (c)	98. (c)	99. (c)	100. (b)
101. (a)	102. (a)	103. (a)	104. (d)	105. (a)
106. (b)	107. (d)	108. (d)	109. (b)	110. (d)
111. (b)	112. (a)	113. (d)	114. (d)	115. (c)
116. (a)	117. (c)	118. (b)	119. (a)	120. (a)
121. (b)	122. (c)	123. (a)	124. (c)	125. (a)
126. (d)	127. (d)	128. (b)	129. (a)	130. (a)
131. (c)	132. (c)	133. (b)	134. (a)	135. (c)
136. (a)	137. (b)	138. (c)	139. (c)	140. (a)
141. (a)	142. (b)	143. (a)	144. (d)	145. (a)
146. (b)	147. (c)	148. (a&b)	149. (d)	150. (c)
151. (a)	152. (d)	153. (c)	154. (d)	155. (a)
156. (b)	157. (c)	158. (b)	159. (c)	160. (a)
161. (d)	162. (d)	163. (d)	164. (b)	165. (c)

166. (d)	167. (a)	168. (d)	169. (a)	170. (a)
171. (c)	172. (a)	173. (a)	174. (a)	175. (a)
176. (d)	177. (c)	178. (d)	179. (b)	180. (a)
181. (c)	182. (b)	183. (c)	184. (c)	185. (c)
186. (b)	187. (b)	188. (a)	189. (c)	190. (d)
191. (d)	192. (c)	193. (b)	194. (d)	195. (c)
196. (d)	197. (b)	198. (c)	199. (b)	200. (b)
201. (c)	202. (c)	203. (a)	204. (a)	205. (b)
206. (b)	207. (d)	208. (a)	209. (a)	210. (a)
211. (c)	212. (d)	213. (d)	214. (d)	215. (c)
216. (d)	217. (a)	218. (b)	219. (c)	220. (a)
221. (a)	222. (a)	223. (b)	224. (a)	225. (b)
226. (d)	227. (a)	228. (b)	229. (b)	230. (a)
231. (c)	232. (c)	233. (a)	234. (c)	235. (d)
236. (a)	237. (b)	238. (d)	239. (a)	240. (b)
241. (b)	242. (a)	243. (a)	244. (a)	245. (b)
246. (d)	247. (c)	248. (a)	249. (b)	250. (d)
251. (a)	252. (d)	253. (d)	254. (d)	255. (b)
256. (b)	257. (a)	258. (d)	259. (a)	260. (a)
261. (b)	262. (c)	263. (a)	264. (c)	265. (d)
266. (d)	267. (d)	268. (d)	269. (a)	270. (b)
271. (d)	272. (c)	273. (a)	274. (d)	275. (a)
276. (d)	277. (b)	278. (d)	279. (b)	280. (a)
281. (d)	282. (c)	283. (a)	284. (c)	285. (a)
286. (b)	287. (b)	288. (c)	289. (d)	290. (c)
291. (a)	292. (b)	293. (c)	294. (c)	295. (b)
296. (c)	297. (b)	298. (d)	299. (a)	300. (c)
301. (b)	302. (c)	303. (d)	304. (c)	305. (a)
306. (c)	307. (a)	308. (a)	309. (b)	310. (d)
311. (a)	312. (b)	313. (d)	314. (d)	315. (b)
316. (a)	317. (c)	318. (b)	319. (a)	320. (a)
321. (d)	322. (d)	323. (b)	324. (c)	325. (a)
326. (d)	327. (b)	328. (c)	329. (d)	330. (b)
331. (b)	332. (a)	333. (b)	334. (c)	335. (d)
336. (b)	337. (a)	338. (b)	339. (d)	340. (c)
341. (a)	342. (a)	343. (d)	344. (c)	345. (b)

346. (c)	347. (a)	348. (d)	349. (b)	350. (d)
351. (b)	352. (c)	353. (a)	354. (b)	355. (c)
356. (d)	357. (a)	358. (b)	359. (d)	360. (c)
361. (b)	362. (c)	363. (c)	364. (b)	365. (c)
366. (c)	367. (b)	368. (d)	369. (d)	370. (a)
371. (b)	372. (d)	373. (b)	374. (c)	375. (d)
376. (c)	377. (d)	378. (a)	379. (b)	380. (c)
381. (c)	382. (b)	383. (c)	384. (c)	385. (d)
386. (a)	387. (c)	388. (a)	389. (b)	390. (a)
391. (d)	392. (c)	393. (b)	394. (d)	395. (c)
396. (c)	397. (a)			

7

ATOMIC AND MOLECULAR SPECTROSCOPY

FORMULAE

1. The equation of wave motion
$$Y = A\sin\omega t$$
   Here,
   Y = displacement
   A = amplitude
   ω = frequency (angular)

2. Velocity of light
$$c = v\lambda \text{ m/s}$$
   Here,
   c = velocity of light
   v = frequency of the light
   λ = wavelength of the e.m.w.

3. Wave number
$$\bar{v} = \frac{1}{\lambda} \text{ cm}^{-1}$$

4. Beer–Lambert law
$$\frac{I}{I_0} = e \times p(-KCl)$$

Here,

I_o = incident intensity of light
I = transmitted intensity of e.m.w.
K = a constant
C = concentration
l = path length

5. The three directions of rotation of a molecule are
   (a) about the bond axis
   (b) end-over-end rotation in the plane of the paper
   (c) end-over-end rotation at right angles to the plane

6. Moment of inertia of polyatomic molecules
   (a) Linear molecule
   $$I_B = I_C \quad I_A = 0$$
   (b) Symmetric tops
   $$I_B = I_C \neq I_A \quad I_A \neq 0$$
   (c) Spherical tops
   $$I_A = I_B = I_C$$
   (d) Asymmetric tops
   $$I_A \neq I_B \neq I_C$$

7. Moment of inertia of diatomic linear molecule
   $$I = \mu r^2$$
   Here, μ = reduced mass.
   $$\frac{m_1 m_2}{m_1 + m_2}$$
   where, m_1, m_2 are the masses of the atoms.

8. Energy level of the molecule
   $$E_J = \frac{h^2}{8\pi^2 I} J(J+1) \text{ joules}$$
   Here,
   I = moment of inertia
   J = rotational quantum number = 0, 1, 2, ...

In terms of wave number

$$\Sigma_J = \frac{E_J}{hc}$$

$$= \frac{h}{8\pi^2 Ic} J(J+1) \text{ cm}^{-1}$$

$$= BJ(J+1)$$

where, B is the rotational constant.

9. The general law of transition

$$\bar{v}J \to J+1 = 2B(J+1) \text{ cm}^{-1}$$

Selection rule $\Delta J = \pm 1$

10. Maximum population at level J

$$J = \sqrt{\frac{KT}{2hcB}} - \frac{1}{2}$$

Here,

K = Boltzmann constant
T = temperature
B = rotational constant

11. Effect of isotope in energy bands

$$\frac{B}{B'} = \frac{\mu'}{\mu}$$

Here,

B, B' = rotational constants
μ, μ' = moment of inertia of corresponding molecules

12. Energy level of non-rigid rotator

$$E_J = \frac{h^2}{8\pi^2 I} J(J+1) - \frac{h^4}{32\pi^4 I^2 r^2 K} J^2(J+1)^2 \text{ Joules}$$

Here,

h = Planck's constant
I = moment of inertia
J = rotational quantum number
r = distance between the molecules

13. The expression for the transitions

$$\Sigma_{J+1} - \Sigma_J = 2B(J+1) - 4D(J+1)^3 \text{ cm}^{-1}$$

where, $D = \dfrac{16B^3 \pi^2 \mu c^2}{K}$.

14. The energy level of simple harmonic oscillator is

$$E_V = \left(V + \dfrac{1}{2}\right) h \omega_{osc} \text{ Joule}$$

Here,
V = vibrational quantum number
ω_{osc} = frequency

In terms of wave number

$$\Sigma_v = \dfrac{E_V}{hc} = \left(V + \dfrac{1}{2}\right) \bar{\omega}_{osc} \text{ cm}^{-1}$$

Here, $\bar{\omega}_{osc}$ = wave number.

15. Morse equation

$$E = D_{eq}[1 - \exp\{a(r_{eq} - r)\}]^2$$

Here,
a = constant for a particular molecule
D_{eq} = dissociation energy

16. The energy of Fundamental absorption

$$\Delta \varepsilon = \bar{\omega}_e (1 - 2x_e) \text{ cm}^{-1}$$

First overtone

$$\Delta \varepsilon = 2\bar{\omega}_e (1 - 3x_e) \text{ cm}^{-1}$$

Hot bands

$$\Delta \varepsilon = 3\bar{\omega}_e (1 - 4x_e) \text{ cm}^{-1}$$

where,
$\bar{\omega}_e$ = equilibrium oscillation frequency
x_e = anharmonicity constant

17. The diatomic vibrating rotator

$$\varepsilon_{total} = \varepsilon_{J,V} = BJ(J+1) + \left(v+\frac{1}{2}\right)\bar{\omega}_e - x_e\left(v+\frac{1}{2}\right)^2 \bar{\omega}_e$$

Here,
- B = rotational constant
- v = vibrational quantum number
- $\bar{\omega}_e$ = equilibrium oscillation frequency
- x_e = anharmonicity constant

18. The selection rules of vibrational and rotational quantum number

$$\Delta V = \pm 1, \pm 2, \text{ etc. } \Delta J = \pm 1$$

19. Fundamental vibrations of a non-linear molecule is

$$3N - 6 \text{ (fundamental vibrations)}$$

For linear molecules,

$$3N - 5 \text{ (vibrations)}$$

where, N = number of atoms.

20. The induced dipole

$$\mu = \alpha E$$

Here,
- α = polarizability
- E = applied electric field

The induced dipole when the electric field $E = E_o \sin 2\pi vt$

$$\mu = \alpha_o E_o \sin 2\pi vt + \frac{1}{2}BE_o\left[\cos 2\pi(v-v_{vib})t - \cos 2\pi(v+v_{vib})t\right]$$

where, $v \pm v_{vib}$ = frequency component of oscillating dipole.

21. Pure rotational Raman spectra
The energy levels

$$\varepsilon_J = BJ(J+1) - DJ^2(J+1)^2 \text{ cm}^{-1}$$

where the selection rule of J is

$$\Delta J = \pm 2$$

22. Zeeman effect

The magnetic dipole of the atom

$$\mu = \frac{-ge}{2m} J$$

$$= \frac{-ge}{2m} \sqrt{j(j+1)}\hbar \ JT^{-1}$$

g is Lande splitting factor

$$= \frac{3}{2} + \frac{S(S+1) - L(L+1)}{2J(J+1)}$$

23. The magnetic interaction energy is

$$\Delta E = \mu_z B_z = \frac{-heg}{4\pi m} B_z j$$

where, B_z = strength of applied magnetic field.

24. Spin angular momentum $I = \sqrt{I(I+1)}\,\hbar$

Here, I = spin quantum number for the nucleus.
Interaction of spin and magnetic field

$$\mu = \frac{q}{2m} I$$

$$= \frac{g\sqrt{I(I+1)}}{2m} \cdot \hbar$$

$$= \frac{qh}{2 \cdot 2\pi m} \sqrt{I(I+1)}\,\hbar \cdot Am^2$$

Here,
q = charge of the electron
m = mass of the electron
μ = dipole

$$\mu = -g\beta \sqrt{I(I+1)} \ JT^{-1}$$

β = Bohr magneton
I = spin quantum number

25. The interaction energy

$$= \mu_z B_z$$

$$\Delta E = \left| E_{I_z} - E_{(I_z-1)} \right|$$

$$= \left| g\beta_N I_z B_z - g\beta_N (I_z - 1) B_z \right|$$

$$= \left| g\beta_N B_z \right| J$$

The frequency emitted

$$\frac{\Delta E}{h} = \left| \frac{g\beta_N B_z}{h} \right| H_z$$

26. Population of energy level

$$\frac{N_{upper}}{N_{lower}} = \exp\left(\frac{-\Delta E}{KT} \right)$$

Here,
ΔE = energy difference
K = Boltzmann constant
T = Temperature

27. The Larmor precession frequency

$$\omega = \frac{\text{Magnetic moment}}{\text{Angular moment}} \times B_z \text{ rad s}^{-1}$$

$$= \frac{\mu B_z}{2\pi I} H_z$$

28. The chemical shift
    The effective magnetic field

$$B_{\text{effective}} = B_{\text{applied}} - B_{\text{induced}}$$

$$B_{\text{induced}} = \sigma B_{\text{applied}}$$

where,
σ = constant

$B_{\text{effective}} = B_0(1-\sigma)$ where, B_0 = applied magnetic field.

29. The energy level separation on applying magnetic field is
$$\Delta E = g\beta B_o$$
Here,
- β = Bohr magneton
- B_o = applied magnetic field
- g = Lande splitting factor

30. Mossbauer spectroscopy

The frequency shift $\Delta v = \dfrac{v \cdot V}{c}$ Hz

Here,
- v = frequency of the light
- c = velocity of the light
- V = relative velocity of source and observer

31. Chemical shift
$$= \text{constant} \cdot \left(\rho_{ex}^2 - \rho_{gd}^2\right) \left(\dfrac{R_{ex} - R_{gd}}{R_{gd}}\right)$$

where,
- ρ_{ex}^2, ρ_{gd}^2 = S electron density of excited and ground state
- R_{ex}, R_{gd} = radius of nucleus when electrons are in excited and ground state

32. Allowed vibrational energy levels of anharmonic oscillator
$$\varepsilon_V = \left(V + \dfrac{1}{2}\right)\bar{\omega}_e - \left(V + \dfrac{1}{2}\right)^2 \bar{\omega}_e x_e \text{ cm}^{-1}$$

Here,
- V = vibrational quantum number
- $\bar{\omega}_e$ = vibrational wave number
- x_e = anharmonicity constant

33. Zero point energy of anharmonic oscillator
$$\varepsilon_0 = \dfrac{1}{2}\bar{\omega}_e\left(1 - \dfrac{1}{2x_e}\right) \text{ cm}^{-1}$$

34. Maximum intensity of transition

$$\bar{v}_{\text{max·int}} = \bar{\omega}_0 \pm 2B\left(\sqrt{\frac{KT}{2Bhc}} + \frac{1}{2}\right)$$

where, positive and negative signs refer to the P and R branches
$\bar{\omega}_0$ = wave number of central frequency
K = Boltzmann constant
T = temperature
c = velocity of light

35. Vibrational Raman spectra

$$\Delta\varepsilon_{\text{fundamental}} = \bar{\omega}_e(1 - 2x_e) \text{ cm}^{-1}$$

$$\Delta\varepsilon_{\text{overtone}} = 2\bar{\omega}_e(1 - 3x_e) \text{ cm}^{-1}$$

$$\Delta\varepsilon_{\text{hot}} = \bar{\omega}_e(1 - 4x_e) \text{ cm}^{-1}$$

Vibrational rotation energy levels

$$\varepsilon_V = \bar{\omega}_e\left(V + \frac{1}{2}\right) - \bar{\omega}_e x_e\left(V + \frac{1}{2}\right)^2 + BJ(J+1) \text{ cm}^{-1}$$

36. Electronic spectra of diatomic molecule
Born–Oppenheimer approximation
$$E_{\text{total}} = E_{\text{ele}} + E_{\text{vib}} + E_{\text{rota}}$$

MULTIPLE CHOICE QUESTIONS

1. In Compton effect
   (a) scattering of the photon and recoil of electron take place simultaneously
   (b) electron recoils after scattering of photon
   (c) scattering follows recoil of electron
   (d) there is no correlation between the two events

2. Which quantum number determines the Stark effect in hydrogen?
   (a) $n, 1, s, j$
   (b) $n, 1, s$
   (c) $n, 1$
   (d) n only

3. What is the value of the Lande g factor for an energy state with $L = 1$ and $J = \frac{3}{2}$?

(a) $\frac{2}{3}$ (b) 1 (c) 2 (d) $\frac{4}{3}$

4. Why is an ammonia maser not useful in communication?
   (a) Because it has low power
   (b) Because its frequency is not convenient
   (c) Because the frequency is not stable
   (d) Because it is not continuous

5. In Rutherford scattering, the distance of closest approach between the incident particle and the target is

(a) $\frac{\alpha}{E}$ (b) $\frac{\alpha}{2E}$ (c) $\frac{2\alpha}{E}$ (d) $\frac{\sqrt{2\alpha}}{E}$

6. The fine structure constant has a value close to

(a) $\frac{1}{11}$ (b) $\frac{1}{111}$ (c) $\frac{1}{137}$ (d) $\frac{1}{115}$

7. The 623.8 nm radiation emitted by a He–Ne laser is due to transition between 3.
   (a) 3s and 2p levels of Ne
   (b) 3s and 3p levels of Ne
   (c) 2p and 2s levels of Ne
   (d) 2p and 1s levels of Ne

8. The Stern–Gerlach experiment showed that the magnetic moment of D atom is (in Bohr magnetons)

(a) 0 (b) $\frac{1}{3}$ (c) $\frac{1}{2}$ (d) 1

9. If σ_{PE} and σ_{PP} are the photoelectric and pair production cross sections for gamma ray scattering, the ratio $\frac{\sigma_{PE}}{\sigma_{PP}}$ is proportional to

(a) Z (b) Z^2 (c) Z^3 (d) Z^4

10. Which of the following is the threshold process?
    (a) Thomson scattering
    (b) photoelectric scattering
    (c) Compton scattering
    (d) pair production

11. The absorption of infrared radiation by an ionic solid like nail is due to
    (a) optical phonons   (b) longitudinal phonons
    (c) impurities   (d) dislocations

12. Under the effect of external field, the $n = 2$ state for hydrogen splits into
    (a) 3 levels   (b) 2 levels
    (c) No splitting   (d) 4 levels

13. The attenuation of intensities observed in rotation Raman spectrum of diatomic molecule is due to
    (a) different Boltzmann factors for the ortho and para states
    (b) different transition probability for the two states
    (c) different nuclear statistical weights for the two states
    (d) interaction between vibrational and rotational states

14. Let P_3 and P_4 represent the optical pumping powers required for the same laser power outputs from a 3-level and 4-level laser. Then
    (a) $P_3 > P_4$ (b) $3P_3 = 4P_4$
    (c) $4P_3 = 3P_4$ (d) $\sqrt{3}P_3 = 2P_4$

15. Consider the fringe pattern obtained on a screen due to diffraction at a single slit. At the position of the first minimum, the phase difference between the wavelets from the opposite edges of the slit is
    (a) $\dfrac{\pi}{2}$ (b) π (c) 2π (d) zero

16. In Compton effect experiment, photons of energy $h\upsilon$ are incident on a target material of atomic number Z. The change in wavelength can be seen more easily if
    (a) υ is in visible region and Z is small
    (b) υ is in X-ray region and Z is small
    (c) υ is in X-ray region and Z is large
    (d) υ is in visible region and Z is large

17. The probability of electrons being captured by the nucleus is maximum for
    (a) K shell electrons
    (b) L shell electrons

(c) M shell electrons

(d) Electrons in the outermost orbits, independent of which shell they come from

18. An unpolarized beam of 100 keV neutrons (de Broglie wavelength $\simeq 20$Å) collides with a stationary proton target. The range of interaction is about 1 fm (10^{-15} m).
    (a) A proton, antiproton or pairs may be produced in the collision.
    (b) The differential scattering cross section will be approximately isotropic.
    (c) The differential scattering cross section will exhibit a significant angular dependence.
    (d) Energy of the scattered neutrons will be unchanged during collisions.

19. Silicon ($Z = 14$) has two electrons in the unfilled $3p$ shells. According to Hund's rule the ground state of Si is
    (a) $^1P_1$
    (b) $^3S_1$
    (c) $^3D_3$
    (d) $^3D_1$

20. An electron is in the $5^2P_{\frac{1}{2}}$ state of the hydrogen atom. It can make an electric dipole transition to
    (a) $2^2P_{\frac{1}{2}}$
    (b) $3^2D_{\frac{5}{2}}$
    (c) the ground state
    (d) $4^2F_{\frac{5}{2}}$

21. There is no infrared absorption for nitrogen molecule because
    (a) its polarizability is zero
    (b) it has no vibrational levels
    (c) it has no rotational levels
    (d) its dipole moment is zero

22. The vibrational constants ω_0 and $\omega_0 x_0$ of HCl are 2937.5 cm^{-1} and 51.6 cm^{-1}. The first Raman Stokes lines will be observed at (in cm^{-1})
    (a) 2989.1
    (b) 2885.9
    (c) 2834.3
    (d) 3040.7

23. The intensity of electronic 0–0 band of a diatomic molecule is very intense when the minimum of the potential curve for the upper electronic state lies (here r is the internuclear distance)
    (a) at the same value of r as that of the lower potential curve
    (b) at a smaller value of r than that of the lower potential curve
    (c) at a large value of r than that of the lower potential curve
    (d) above the dissociation level of the lower potential curve

24. The frequencies of lines of a line spectrum of X-ray emission depends on
    (a) the kinetic energy of the electron
    (b) the metal used for the anticathode
    (c) the deceleration of the electron
    (d) the shape of the continuous spectrum

25. The dispersion relation of a certain wave is $\omega = \sqrt{c^2 k^2 + m^2}$ where ω is the angular frequency, k is the wave vector, c is the velocity of light and m is a constant. The group velocity v of the wave has the following properties
    (a) $v \to c$ as $k \to 0$ and $v \to c$ as $k \to \infty$
    (b) $v \to c$ as $k \to 0$ and $v \to \infty$ as $k \to \infty$
    (c) $v \to 0$ as $k \to 0$ and $v \to \infty$ as $k \to \infty$
    (d) $v \to 0$ as $k \to 0$ and $\omega \to c$ as $k \to \infty$

26. Sodium atom has 11 electrons. If the sequence in which the energy levels are filled is 1s, 2s, 2p, 3s, 3p, 4s, 3d,... the ground state of sodium is
    (a) $^3P_{1/2}$ (b) $^2P_{1/2}$ (c) $^1P_{1/2}$ (d) $^2S_{1/2}$

27. The number of photons emitted per second from a 1-watt Ar-ion laser operating at 488 nm is approximately
    (a) 10.23×10^{19} (b) 2.46×10^{18}
    (c) 10.23×10^{17} (d) 2.46×10^{15}

28. Spectral linewidth of the He–Ne laser is 0.01 nm and the cross sectional area of the beam is 0.01 cm². If the output power is 1 milliwatt, the radiation intensity/unit wavelength (in watt/cm³) is
    (a) 10^{10} (b) 10^8 (c) 10^{-8} (d) 10^{-10}

29. The maximum kinetic energy of photoelectrons in a photoelectric effect depends on
    (a) intensity of incident light
    (b) frequency of the incident light
    (c) polarization of the incident light
    (d) angle of incidence

30. In Young's double slit experiment with a Helium–Neon laser beam of wavelength 632 nm, the first interference maximum will occur when
    (a) path difference is 948 nm
    (b) path difference is 316 nm
    (c) phase difference is 2π radians
    (d) phase difference is π radians

31. Electromagnetic radiations will be emitted in the case of a
    (a) neutron moving in a straight line with a constant speed
    (b) proton
    (c) proton moving in a circle with a constant speed
    (d) electron moving in a straight line with a constant speed

32. Typical energy of the rotational modes in a polyatomic molecule like NH_3 is
    (a) 10^6 eV (b) 10^{-3} eV (c) 10^{-1} eV (d) 1 eV

33. The energy separation between two consecutive Stokes line in Raman scattering depends on
    (a) energy separation between vibrational levels in the excited state
    (b) wavelength of the incident light
    (c) energy separation between vibrational levels in the ground state
    (d) intensity of incident light

34. A line arising from a transition of an atom from an excited state to the ground state has a width of 0.5 cm^{-1}. If the width is almost entirely due to the width of the initial state, the life time of the state will be closest to
    (a) 6×10^{-12} s   (b) 6×10^{-15} s
    (c) 6×10^{-9} s   (d) 6×10^{-17} s

35. In the de Broglie hypothesis, waves are
    (a) associated with particle at rest
    (b) associated with a particle in motion
    (c) associated with a medium
    (d) not at all involved

36. The double-slit electron interference pattern appears when a large number of electrons are accumulated at the screen. According to the quantum theory, the interference pattern appears
    (a) only if a beam of electrons are in transit so that one part of the beam goes through one slit and another part through the other
    (b) even if one electron at a time is in transit, but an electron goes through one of the two slits only
    (c) even if one electron at a time is in transit, but an electron goes through both the slits at once
    (d) the consideration of an electron at a time in transit is not physically possible

37. Which of the following represents the electro-optic effect?
    (a) Zeeman effect   (b) Paschen–Back effect
    (c) Compton effect   (d) Stark effect

38. The separation in which the energy levels are equally spaced is
    (a) vibrational spectra
    (b) rotational spectra
    (c) rotational-vibrational spectra
    (d) electron spectra

39. In an electron band system resulting from a transition between states with nearly equal vibrational constants, the intense sequence correspond to
    (a) $\Delta V = 0$ sequence   (b) $\Delta V = 1$ sequence
    (c) $\Delta V = \pm 1$ sequence   (d) $\Delta V =$ both $+1$ and -1 sequence

40. Using a Bragg's spectrometer, the glancing angle for the first-order spectrum was observed to be so if $d = 2.8 \text{Å}$, the wavelength of X-rays is
    (a) 0.79Å (b) 0.39Å (c) 0.5Å (d) 0.25Å

41. The intercept of the polarizability versus temperature $\left(\dfrac{1}{T}\right)$ curve on the polarizability axis gives
    (a) sum of orientational and electric parts
    (b) sum of ionic and orientational components
    (c) sum of ionic and electric components
    (d) only orientational component

42. The Rutherford scattering cross section is proportional to $\operatorname{cosec}^{n}\dfrac{\psi}{2}$ where ψ is the scattering angle and the index n is equal to
    (a) 1   (b) 2   (c) 3   (d) 4

43. The cause of the 'Red Shift' in the photon frequency of a source is
    (a) longitudinal Doppler effect
    (b) Compton effect
    (c) transverse Doppler effect
    (d) fluorescence

44. A beam of plane polarized light passes through a medium held in a current-carrying solenoid and undergoes a 'Faraday rotation'. If the beam is reflected back at the end of the solenoid, what is the rotation at the starting end?
    (a) 2θ (b) θ (c) 0   (d) θ^2

45. Which of the following shows optical activity?
    (a) common salt solution   (b) area solution
    (c) sugar solution   (d) water

46. According to the optic theorem at limiting distance
    (a) the dimension of scatterer and shadow are independent of each other
    (b) the dimension of the shadow is very much larger than that of the scatterer
    (c) the dimension of the shadow is very much smaller than that of the scatterer
    (d) the dimension of the shadow and scatterer are of the same order

47. In the first order Stark effect in hydrogen atom, the ground state
    (a) splits into two levels   (b) splits into three levels
    (c) splits into four levels   (d) does not split

48. The magnetic moment (in units of Bohr magnetons) of the electron due to its spin is

(a) $\dfrac{1}{2}$ (b) 1 (c) 1.25 (d) 1.33

49. The fine structure constant α occurring in the relativistic correction for the energy of a hydrogen atom is given by

(a) $\dfrac{2\pi e^2}{hc}$ (b) $\dfrac{\pi e^2}{hc}$ (c) $\dfrac{\pi e^2}{2hc}$ (d) $\dfrac{2\pi e^2}{hmc}$

50. In the presence of a magnetic field, a state with total angular momentum quantum number J splits into
   (a) J components
   (b) $2J$ components
   (c) $2J + 1$ components
   (d) $2(J + 1)$ components

51. The alternation of intensities observed in the pure rotational Raman spectrum of homonuclear diatomic molecules is due to
   (a) isotope effect
   (b) nuclear spin
   (b) electron spin
   (d) Doppler effect

52. The red light emitted by a He–Ne laser is due to transitions between
   (a) He energy levels
   (b) He and Ne energy levels
   (c) Ne energy levels
   (d) impurity energy levels

53. Which of the following statements about the magnetic moment μ of an atom is correct? Here, L and S denote the total orbital angular momentum and spin angular momentum.

(a) $\mu = \dfrac{e}{2mc} \times S$ (b) $\mu = \dfrac{e}{2mc}(L+S)$

(c) $\mu = \dfrac{e}{2mc}(L+2S)$ (d) $\mu = \dfrac{eh}{2mc}gh$

$1 < g < 2$

54. The Lande's splitting factor for the atomic state $^2P_{\frac{3}{2}}$ is

(a) $\dfrac{1}{3}$ (b) $\dfrac{2}{3}$ (c) 1 (d) $\dfrac{4}{3}$

55. Antiparallel ordering of spins in crystals can be detected by
   (a) electron diffraction
   (b) electron microscope
   (c) neutron diffraction
   (d) X-ray diffraction

56. The plane of polarization of a light beam travelling from one end of a polaroid to the other end undergoes a rotation θ. If the beam is now reflected and reaches the original end along the same path, the resultant rotation is

(a) 0 (b) θ (c) 2θ (d) θ^2

57. Attenuation of intensities is observed in the rotation-vibration spectrum of

(a) C_2H_2 (b) HBr (c) CO_2 (d) N_2O

58. Which of the following is optically active?

(a) stretched polymer sheet (b) calcite
(c) silvered glass (d) quartz

59. In the He–Ne laser the role of the He atom is

(a) to emit the red light
(b) to control the output
(c) to control the wavelength
(d) to effect population inversion between the Ne levels

60. The spectroscopic phenomenon which led to the discovery of heavy hydrogen is

(a) Zeeman effect
(b) Stark effect
(c) nuclear spin hyperfine structure
(d) isotope effect

61. If V_p and V_g are the phase and group velocities of electromagnetic waves passing through a medium, then in the region of anomalous dispersion

(a) $V_p > C, V_g > C$ (b) $V_p < C, V_g > C$
(c) $V_p > C, V_g < C$ (d) $V_p < C, V_g < C$

62. The transverse Doppler effect can be explained in terms of

(a) the length contraction (b) the time dilation
(c) both (a) and (b) (d) neither (a) nor (b)

63. When an electron in a hydrogen atom moves from an orbit to another orbit of longer radius, which of the following decreases?

(a) potential energy (b) total energy
(c) angular momentum (d) rotational speed

64. With what aspects of radiation is the equation $\log \lambda = \log R + B$ associated?
   (a) characteristic X-rays
   (b) black body radiation
   (c) α-emission
   (d) Compton effect

65. In Rutherford's scattering formula, the scattering cross section for angle θ is proportional to
   (a) $\sin^4\left(\dfrac{\theta}{2}\right)$
   (b) $\sin^{-4}\left(\dfrac{\theta}{2}\right)$
   (c) $\sin^2\left(\dfrac{\theta}{2}\right)\cos^2\left(\dfrac{\theta}{2}\right)$
   (d) $\cot^4\left(\dfrac{\theta}{2}\right)$

66. Which of the following experiment involves the four concepts of discrete energy levels—Larmor precession, space quantization and L-S coupling?
   (a) Paschen–Back effect
   (b) Frank–Hertz experiment
   (c) Stern and Gerlach experiment
   (d) Zeeman effect

67. The transition for the sodium D_1 line is
   (a) $^2P_{\frac{3}{2}} - {}^2S_{\frac{1}{2}}$
   (b) $^2P_{\frac{1}{2}} - {}^2S_{\frac{1}{2}}$
   (c) $^2D_{\frac{3}{2}} - {}^2P_{\frac{1}{2}}$
   (d) $^2D_{\frac{5}{2}} - {}^2P_{\frac{3}{2}}$

68. What is the effective mass of a photon of wavelength $1\,\text{Å}$?
   (a) 4.42×10^{-36} kg
   (b) 2.208×10^{-36} kg
   (c) 2.208×10^{-32} kg
   (d) 4.42×10^{-32} kg

69. In which of the following do all spectral lines show a triplet structure?
   (a) Zeeman effect
   (b) Paschen–Back effect
   (c) Stark effect
   (d) hyperfine structure

70. As the temperature is increased, the intensity of an anti-Stokes Raman lines
   (a) increases
   (b) decreases

(c) remains unchanged

(d) increases and decreases depending on the mode of vibration

71. The intensities of lines in the rotational Raman of H_2 molecule are in the ratio
    (a) 1 : 1   (b) 1 : 2   (c) 1 : 3   (d) 1 : 4

72. The Rutherford scattering cross section is
    (a) directly proportional to the energy of the incident particle
    (b) inversely proportional to the square of the energy of the incident particle
    (c) inversely proportional to the sine of the scattering angle
    (d) directly proportional to the product of the charges of the projectile and target particles

73. By hyperfine structure, we mean
    (a) energy level splitting of vibrational spectra due to rotation of molecules
    (b) energy level splitting of electron spectra due to vibration of molecules
    (c) energy level splitting of electronic spectra due to nuclear spin
    (d) energy level splitting of vibrational spectra due to spin of the electron

74. The spectroscopic term for a perpendicular state is quoted as $^2S_1$. If g is the Lande's g-factor then
    (a) No such state can exist
    (b) $L = 0; S = \frac{1}{2}; J = \frac{1}{2}; g = 2$
    (c) $L = 0; S = \frac{1}{2}; J = 1; g = 2$
    (d) $L = 2; S = 0; J = 1; g = 1$

75. Equivalent electrons have
    (a) the same j only   (b) the same n only
    (c) the same m only   (d) none of the above

76. Choose the wrong statement concerning Zeeman effect.
    (a) It gives useful information about the rotational and vibrational states of molecules.
    (b) It is the lifting of degeneracy of energy levels on the application of an external magnetic field.

(c) The term symbols for various atomic states can be deduced by the Zeeman effect.

(d) Normal Zeeman effect applies to transitions between singlet states only.

77. According to Franck–Condon principle,
   (a) an electronic transition takes place so rapidly that a rotating molecule does not change its internuclear distance appreciably during the transition
   (b) the vibrational lines in a progression are all of the same intensity
   (c) the intensity variation of rotational-vibrational spectra of molecules can be explained using the Morse curve
   (d) a vibrating molecule does not change its internuclear distance appreciably during an electronic transition

78. Choose the correct answer concerning the He–Ne laser.
   (a) It is an example of a pulsed laser and a three-level system operating as a laser.
   (b) The electric discharge essentially pumps helium atoms, in a mixture of helium and neon into an excited state which undergoes induced decay.
   (c) If radiation of about 8700 cm^{-1} is present, the laser radiation emitted has a wavelength of 632.8 nm.
   (d) The presence of radiation of about 15800 cm^{-1} gives rise to the 632.8 nm radiation from this laser.

79. Which one of the following quantities represents $\frac{hc}{2\pi}$?
   (a) 1973 eV/Å
   (b) 973 Å/eV
   (c) 938 MeV/Å
   (d) 1.973 keV/Å

80. An optical plane wave is incident normally on one face of a rectangular glass slab whose refractive index increases linearly from 1.5 to 1.8 from side P to side Q. After refraction, the wave will

   (a) go undeviated
   (b) be deviated towards side P

(c) be deviated towards side Q
(d) be totally reflected

81. The hyperfine splitting of atomic levels is due to
    (a) spin–orbit interaction
    (b) electron spin–nuclear spin interaction
    (c) electron spin–electron spin interaction
    (d) relativistic correction

82. Let A and B be the Einstein spontaneous and stimulated emission coefficients respectively for a pair of levels separated by energy $h\nu$. The ratio of A/B
    (a) does not depend on ν
    (b) is proportional to ν
    (c) is proportional to ν^{-2}
    (d) is proportional to ν^{-3}

83. Electron spin resonance can be exhibited by
    (a) hydrogen atom
    (b) hydrogen molecule
    (c) Li^+ ion
    (d) nitric oxide (NO) molecule

84. A spin $\frac{3}{2}$ atom is subjected to a constant external electric field. The possible schematic energy level splitting/splittings is/are

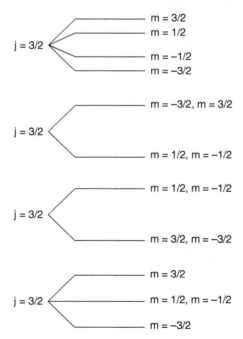

85. The linear Stark effect is possible in a hydrogen atom but not in a sodium atom because
    (a) the principle quantum for the ground state of the sodium atom is different from that of the hydrogen atom in the ground state
    (b) spin-orbit interaction is stronger in sodium than in hydrogen
    (c) the electronic energy levels of sodium exhibit l-degeneracy
    (d) the electronic energy levels of hydrogen exhibit l-degeneracy

86. A stable H molecule is possible because
    (a) the electronic ground state is a spin triplet
    (b) the magnitude of the electronic charge density is maximal in the internuclear region
    (c) the electronic charge density is depleted in the internuclear region
    (d) of strong spin-orbit interaction in the molecule

87. An atom with nuclear spin equal to $\frac{1}{2}(\hbar)$ cannot have
    (a) fine structure
    (b) hyperfine structure
    (c) electric quadrupole interaction
    (d) magnetic interaction

88. Consider hydrogen atoms undergoing radioactive transitions while emitting the spectral lines listed below. The shift in frequency due to recoil of the emitted atom will be the largest for
    (a) $Ly\alpha$
    (b) $Ly\beta$
    (c) $H\alpha$
    (d) $H\beta$

89. In a Millikan oil drop experiment, it was found that an oil drop of weight 3.2×10^{-13} N could be balanced (i.e., held stationary) between two parallel metallic plates when the field between the plates was 5×10^5 N/M. The charge on the drop is
    (a) $3e$
    (b) $4e$
    (c) $6e$
    (d) $8e$

90. The minimum photon energy required for the phot-disintegration of the deuteron is about
    (a) 0.51 MeV
    (b) 1.02 MeV
    (c) 2.22 MeV
    (d) 13.6 MeV

91. A thin film of water (refractive index = 4/3) is 3100Å thick. If it is illuminated by white light at normal incidence the colour of the film in reflected light will be
   (a) blue   (b) green   (c) yellow   (d) red

92. The number of electrons in the upper level of lesser materials must be
   (a) higher than the equilibrium population
   (b) higher than the population of the lower level
   (c) lower than the population in the lower level
   (d) equal to lower level

93. The Bragg's angle corresponding to the first order reflection (III) planes in a crystal is 30° when X-ray of wavelength 1.75Å are used. The inter-planar spacing is
   (a) 1.75Å   (b) 17.5Å   (c) 0.017Å   (d) 0.175Å

94. Rayleigh scattering is due to
   (a) fluctuations in absorption
   (b) stimulated emission
   (c) vibrations of O—H bonds
   (d) fluctuations in refractive index

95. Suppose an electromagnetic wave has an electric field amplitude of 10^{-5} V/m. The amplitude of the B wave is
   (a) 3.3×10^{-14} T
   (b) 3.3×10^{-7} T
   (c) 3.0×10^{8} T
   (d) 3.3×10^{-10} T

96. The energy level of an electron in a tube of length 2.0 cm and 200 nm is
   (a) $1.5 \times 10^{-34} n^2$ J and $1.5 \times 10^{-24} n^2$ J
   (b) $3.5 \times 10^{-34} n^2$ J and $3.5 \times 10^{-24} n^2$ J
   (c) $4.0 \times 10^{-10} n^2$ J and $4.0 \times 10^{-6} n^2$ J
   (d) $5 \times 10^{-10} n^2$ J and $5 \times 10^{-16} n^2$ J

97. The index of refraction of benzene is 1.5. The velocity of light in benzene is
   (a) 1.5×10^8 m/s
   (b) 2×10^8 m/s
   (c) 3×10^8 m/s
   (d) 4.5×10^6 m/s

98. At absolute zero may be regarded as that temperature at which
    (a) water freezes
    (b) all gases become liquids
    (c) all substances are solids
    (d) molecular motion in a gas would be the minimum possible

99. If the distance between a proton and an electron is doubled, the resulting attractions will be
    (a) four times as great
    (b) twice as great
    (c) half as great
    (d) one-fourth as great

100. If the speed of light were $\frac{2}{3}$ of its present value, the energy released in a given atomic explosion will be
    (a) decreased by a factor $\frac{2}{3}$
    (b) decreased by a factor $\frac{4}{9}$
    (c) decreased by a factor $\frac{5}{9}$
    (d) decreased by a factor $\frac{\sqrt{5}}{9}$

101. The presence of three unpaired electrons is explained by
    (a) Pauli's exclusion principle
    (b) Aufbau principle
    (c) Uncertainty principle
    (d) Hund's rule

102. Miller effect in a transistor accounts for
    (a) variations in its α
    (b) decreases in its β with frequency
    (c) collector-base feedback
    (d) emitter-base feedback

103. The index of refraction of benzene is 1.5. The velocity of light in benzene is
    (a) 1.5×10^8 m/s
    (b) 2×10^8 m/s
    (c) 3×10^8 m/s
    (d) 4.5×10^8 m/s

104. A diffraction pattern is obtained using a beam of red light. Now red light is replaced by a blue light, then
    (a) the bands will disappear
    (b) diffraction bands will become broader and farther apart

(c) diffraction bands will become narrower and broader together
(d) no change

105. C–H stretching absorbs at higher wave number (IR) compared to C–C band. This is because of the
    (a) smaller value of the reduced mass of O–H compared with C–C
    (b) higher value of the reduced mass of O–H compared with C–C
    (c) higher value of total mass of O–H compared with C–C
    (d) stronger force constant for O–H than C–C

106. Bohr's correspondence principle asserts that the motion of a system as described by quantum mechanics and by classical mechanics must agree in the limit in which the Planck's constant h
    (a) is large
    (b) is small and can be neglected
    (c) is infinite
    (d) none of the above

107. The exchange energy in hydrogen molecule problem is a consequence of
    (a) Pauli's exclusion principle
    (b) Heisenberg's uncertainty principle
    (c) Bohr's correspondence principle
    (d) Einstein's mass energy relation

108. The reduced mass (μ) for a system of two particles is

    (a) $\mu = \dfrac{1}{m_1} + \dfrac{1}{m_2}$
    (b) $\mu = \dfrac{m_1 m_2}{m_1 - m_2}$

    (c) $\dfrac{m_1 + m_2}{m_1 m_2}$
    (d) $\dfrac{1}{\mu} = \dfrac{1}{m_1} + \dfrac{1}{m_2}$

109. The order of colour in a secondary rainbow is
    (a) red on the inner edge and violet on the outer edge
    (b) red on the outer edge and violet on the inner edge
    (c) the regular pattern is observed
    (d) none of the above

110. A man suffering from short sight is unable to see objects distinctly at a distance greater than 3 m. The power of spectacle lenses with which he will able to view distant objects distinctly is

(a) −0.33D (b) +0.33D (c) +6.66D (d) −6.66D

111. To explain fine structure of the spectrum of hydrogen atom we must consider

(a) finite size of the nucleus
(b) the presence of neutrons in the nucleus
(c) spin angular momentum
(d) orbital angular momentum

112. Interplanar molecular distance of sodium chloride is 2.52Å. At what angle must an X-ray beam of wavelength 1.1Å fall on a family of planes of sodium chloride crystal if diffracted beam is to exist?

(a) $\sin^{-1}(0.2183)$ (b) $\sin^{-1}(0.5)$
(c) $\sin^{-1}(0.7)$ (d) $\sin^{-1}(0.9)$

113. If the speed of light were $\frac{2}{3}$ of its present value, the energy released in a given atomic explosion will be

(a) decreased by a factor $\frac{2}{3}$ (b) decreased by a factor $\frac{4}{9}$
(c) decreased by a factor $\frac{5}{9}$ (d) decreased by a factor $\frac{\sqrt{5}}{9}$

114. In Compton effect, photons of wavelength λ and frequency υ scatter at angle ϕ with modified wavelength λ' and frequency υ' which of the following is true?

(a) $\lambda' - \lambda$ varies with ϕ and also with the scatter
(b) $\upsilon - \upsilon'$ is independent of the scattering material
(c) $\lambda' - \lambda$ varies with ϕ in proportion to $(1 + \cos\phi)$
(d) $\lambda' - \lambda$ is independent of the scatterer but varies with ϕ

115. In a laser source the emission from various atoms/molecules are coherent

(a) in phase only
(b) in direction of emission only
(c) in phase, direction and polarization
(d) in phase and polarization only

116. The energy terms for alkali atoms are given by $\dfrac{R}{(n-\alpha)^2}$, where n is the total quantum number and α is different for the s, p, d substates. (For H atom, α is zero for all substates). Which of the following is true?

   (a) α is always negative
   (b) α is larger for p than for s substate
   (c) α varies substantially with n for the same substate for a given atom
   (d) α decreases as we go from s to p to d substate

117. Covalent bonds are formed by crystals of

   (a) carbon   (b) hydrogen
   (c) neon   (d) water

118. Black holes refer to

   (a) holes in the heavenly bodies
   (b) sun spots
   (c) collapsing objects of low density
   (d) collapsing objects of high density

119. The value of Bohr magneton m_B is

   (a) 9.27×10^{-24} J/T
   (b) 92.7×10^{-24} J/T
   (c) 0.927×10^{-24} J/T
   (d) 927×10^{24} J/T

120. Fringes in the Michelson interferometer are circular because

   (a) circular reflectors are used
   (b) they are fringes of equal inclination
   (c) fringes of equal thickness
   (d) light is emitted as spherical waves

121. A spring of force constant K is cut into three equal parts. Then the force constant of each part is

   (a) K (b) $\dfrac{K}{3}$ (c) $3K$ (d) $3K^2$

122. Plasma is a cloud of
    (a) α-particle
    (b) neutrino
    (c) completely ionized matter
    (d) uncharged particles
123. Maser is a device in which
    (a) light waves interact with matter
    (b) microwaves interact with matter
    (c) mesons interact with microwaves
    (d) gamma rays interact with matter
124. In ruby laser, the ruby crystal contains
    (a) 10% of chromium   (b) 0.05% of chromium
    (c) 0.5% of chromium   (d) 0.0005% of chromium

ANSWERS

1. (a)	2. (b)	3. (d)	4. (b)	5. (b)
6. (c)	7. (a)	8. (d)	9. (b)	10. (b)
11. (a)	12. (d)	13. (b)	14. (a)	15. (c)
16. (c)	17. (a)	18. (b)	19. (d)	20. (c)
21. (d)	22. (c)	23. (d)	24. (a)	25. (a)
26. (d)	27. (b)	28. (d)	29. (b)	30. (c)
31. (d)	32. (b)	33. (c)	34. (c)	35. (b)
36. (c)	37. (c)	38. (b)	39. (a)	40. (b)
41. (c)	42. (d)	43. (a)	44. (c)	45. (c)
46. (c)	47. (d)	48. (b)	49. (a)	50. (c)
51. (a)	52. (c)	53. (c)	54. (d)	55. (a)
56. (c)	57. (b)	58. (d)	59. (d)	60. (d)
61. (d)	62. (b)	63. (d)	64. (c)	65. (b)
66. (d)	67. (b)	68. (c)	69. (a)	70. (b)
71. (c)	72. (b)	73. (c)	74. (b)	75. (c)
76. (c)	77. (d)	78. (b)	79. (b)	80. (a)
81. (b)	82. (c)	83. (d)	84. (a)	85. (a)
86. (d)	87. (c)	88. (b)	89. (c)	90. (c)

91. (c)	92. (b)	93. (a)	94. (d)	95. (a)
96. (c)	97. (b)	98. (d)	99. (d)	100. (b)
101. (d)	102. (c)	103. (b)	104. (c)	105. (d)
106. (d)	107. (a)	108. (d)	109. (b)	110. (a)
111. (d)	112. (a)	113. (b)	114. (d)	115. (c)
116. (a)	117. (a)	118. (a)	119. (a)	120. (b)
121. (a)	122. (c)	123. (b)	124. (b)	

8

NUCLEAR PHYSICS AND ELEMENTARY PARTICLES

FORMULAE

1. The force of repulsion between two similar charges by Coulomb's law is

$$F = \frac{ZeE}{r^2}$$

where,
   r = distance of the α-particle from nucleus
   E = charge of α-particle
   Ze = charge of the nucleus

2. Rutherford's scattering formula is

$$N = Q_n t \frac{ds}{4R^2} \left(\frac{ZeE}{Mv^2}\right)^2 \operatorname{cosec}^4 \frac{\theta}{2}$$

Here,
   Q = total number of α-particles
   N = number striking
   E = charge on α-particle
   t = thickness of the foil
   ds = area of the screen
   R = distance of the screen from the point of scattering
   θ = angle of scattering

3. Radius of the nucleus is
$$R = R_0 A^{\frac{1}{3}}$$
where,
R_0 = nuclear unit radius
A = mass number

4. The Fermi model, the model of the nuclear charge distribution, is represented by the equation
$$\rho(r) = \frac{\rho_1}{1 + e^{K(r-c)}}$$
where,
ρ_1, K = constants
c = distance from the centre of the nucleus
α = point where $\rho(r)$ has fallen to half its central value

5. Binding energy is given by
$$E_B = \Delta M c^2$$
$$= \left[Z M_H + (A - Z) M_N - {}_Z M^A \right] c^2$$
Here,
ΔM = mass defect
M_H, M_N = the masses of hydrogen atom and neutron
${}_Z M^A$ = isotropic atomic mass

6. Weizsacker's semi-empirical mass formula is
$${}_Z M^A = Z M_H + (A - Z) M_N - a_v A + a_s A^{\frac{2}{3}}$$
$$+ a_c \frac{Z^2}{Z^{\frac{1}{3}}} + a_a \frac{(A - 2Z)^2}{A} + a_p \frac{1}{A^{\frac{3}{4}}}$$
Here,
a_p = pairing energy
a_v = volume binding energy
a_c = Coulomb energy
a_s = surface binding energy
a_a = asymmetry energy

7. Nuclear magnetic moment vector is given as
$$\mu = g_n \mu_n I$$
Here,
g_n = ratio of nuclear magnetic moment
μ = magnetic moment of nucleus
I = spin of the nucleus

8. Quadrupole moment is given by the formula
$$Q = \frac{e}{2}(3Z^2 - r^2)$$

9. The average kinetic energy for nucleus in a particular nucleus is
$$E = \frac{h^2}{8MR_0^2 A^{\frac{2}{3}}}$$
Here,
R_0 = nuclear unit radius
A = mass number

10. Number of nuclides (radioactive) at any time t in a nucleus is given by
$$N = N_0 e^{-\lambda t}$$
where,
N_0 = total number of nuclides initially
λ = disintegration constant $= \dfrac{0.693}{T}$
T = time taken by the one half of the initial number of nuclei in a sample to disintegrate (half-life period)

11. Geiger law, for ranges in standard air, between three and seven cms, or for the medium range is
$$R = a V_0^3$$
Here,
R = extrapolated range in cms
V_0 = initial velocity in cm/sec
a = constant numerically equal to 9.6×10^{-28}

12. Geiger–Nuttal law, relating the half-life of an α-emitter and range of its α-particles, is given by the formula
$$\log \lambda = A \log R + B$$

Here,
A, B = constants
R = extrapolated range

13. The α-disintegration energy E is given by
$$E = \frac{1}{2}MV^2 + \frac{1}{2}M_R V_R^2$$
Here,
M_R = mass of product nucleus
V_R = velocity of product nucleus

14. The helicity of a particle is defined as
$$H = \hat{\sigma} \cdot \hat{p}$$
where,
$\hat{\sigma}$ = unit vector in spin direction
$\hat{p}$ = unit vector in direction of the linear momentum of the particle

15. The cross section for pair production in the field of the nucleus is
$$\sigma_{pp} = \frac{Z^2}{137}\left(\frac{e^2}{m_o c^2}\right)\left(\frac{28}{9}\log_c \frac{2h\upsilon}{m_o c^2} - \frac{218}{27}\right)$$
Here,
$m_o c^2$ = rest mass
Z = atomic number

16. The counting rate I for monoenergetic gamma radiations is given by the relation
$$I = I_o e^{-\mu\tau}$$
Here,
I_o = counting rate in the absence of absorber
μ = linear absorption coefficient of the absorber material

17. The energy of recoil of the photon is given by
$$E_r = \frac{p^2}{2M}$$
Here,
M = mass of the nucleus

p = momentum of photon $\left(\dfrac{E\hat{e}}{e}\right)$

$E_{\hat{e}}$ = energy of the emitted photon

18. The time taken by the charged particle to traverse the semicircular path in the dee in the cyclotron is given by

$$t = \dfrac{\pi M}{Hq}$$

Here,

H = uniform magnetic field at right angles to the plane of the dees

M = mass of the ion

19. The kinetic energy of the ion emerging from the cyclotron is

$$E = 2\pi^2 R^2 f^2 M$$

Here,

R = radius of the dee

f = frequency of the alternating potential

M = mass of the ion

20. The Lawson criterion for self sustaining thermonuclear system is given by

$$\text{Lawson criterion} = nt$$

Here,

n = number of reacting nuclei per cubic cm

t = time in seconds during which thermonuclear reaction takes place

21. The four-factor formula is given by

$$k_\alpha = \eta \varepsilon p f$$

Here,

k_α = multiplication factor

f = thermal utilization factor

p = resonance escape probability

ε = fast fission factor

η = fast fission neutrons

22. The Fermi age is given by
$$\nabla^2 q(r,\tau) = \frac{\partial q(r,\tau)}{2\tau}$$
Here,
- q = slowing down density function of the space coordinates r and τ
- τ = neutron age

23. The general diffusion equation for an actual reactor is
$$\nabla^2 \phi + \left(\frac{k_\infty e^{-B^2\tau} - 1}{L^2}\right)\phi = 0$$
Here,
- k_∞ = multiplication factor
- B = geometrical buckling
- L = diffusion length of thermal neutrons
- τ = neutron age

24. The cross section for a (n, p) reaction is
$$\sigma(n,p) = (R_0 + \lambda)^2 \frac{\lambda_p}{\lambda_T}$$
$$\lambda_T = \lambda_n + \lambda_p + \lambda_\alpha$$
Here,
- λ_T = total decay rate
- λ_n = total decay rate for neutron emission
- λ_p = total decay rate for proton emission
- λ_α = total decay rate for alpha particle emission

25. The cross section for a (p, ∞) reaction is
$$\sigma(p, \infty) = \sigma_{RP} \frac{\lambda_\infty}{\lambda T'}$$

26. The Breit–Wigner formula is given as
$$\sigma_{(n,x)} = \pi\lambda^2 \frac{\Gamma n \, \Gamma x}{(E - E_\tau)^2 T \frac{1}{4}\Gamma^2}$$
Here,
- $\pi\lambda^2$ = maximum probable cross section
- Γ = total width

27. The Majorana potential V_M, as it operates on a wave function ψ_1 may be defined as
$$V_M\psi = v_M(r)P^x\psi$$
Here,

$v_M(r)$ = ordinary function of r

P^x = operator which exchanges the positions of the two particles in the wave function

28. The Bartlett potential V_B is defined as
$$V_B\psi = v_B(r)P^S\psi$$
Here,

$v_B(r)$ = ordinary function of r

P^S = operator which interchanges the spins

29. The Heisenberg potential V_H is defined as
$$V_H\psi = v_H(r)p^x P^S\psi$$

30. The Wigner force is given by
$$V_\omega\psi = v_\omega(r)\psi$$

31. For a homogeneous thermal reactor, the thermal utilization factor is
$$f = \frac{\Sigma_{aF}}{\Sigma_{aF} + \Sigma_{aM}}$$
Here, Σ_{aF} and Σ_{aM} are the macroscopic thermal absorption cross section of fuel and modulator.

32. For a heterogeneous reactor, thermal utilization is defined as
$$f = \frac{\Sigma_{aF} V_F}{\Sigma_{aF} V_F + \Sigma_{aM} V_M \zeta}$$
Here, ζ = thermal disadvantage factor.

33. In the synchrotron, the circulating frequency at any moment is given by
$$\omega_0 = \frac{B_{ec}}{E}$$
Here, E = total particle energy.

34. The number of interactions for a thin target is given by
$$N_1 = N_p \cdot N_T \cdot \sigma \cdot n$$
Here,
   N_1 = no. of interactions/second
   N_p = no. of incident particles/pulse
   N_T = no. of target particles/unit area
   σ = interaction cross section
   n = no. of pulses/second

35. The number of target particles is given as
$$N_T = \frac{N_A}{A} \cdot \rho \cdot t \cdot N$$
Here,
   N_A = Avogadro's number
   A = atomic weight
   ρ = target density
   t = target thickness
   N = no. of target particles/atom

36. The luminisity in colliding beam machines is given as
$$L = \frac{n_1 n_2}{a} \cdot b \cdot f$$
Here,
   n_1, n_2 = no. of particles/bunch in each beam
   a = cross sectional area of beams at intersection
   b = no. of bunches/beam
   f = revolution frequency

37. For a particle travelling in a circle, the energy loss per turn due to synchrotron radiation is given by
$$\Delta E = \frac{4\pi}{3} e^2 \beta^3 \frac{1}{R} \left(\frac{E}{mc^2}\right)^4$$
Here,
   e = charge
   β = velocity (c)
   E = total energy

m = mass
R = orbit radius

38. The minimum kinetic energy necessary to produce a π-meson is

$$T = m_\pi \left[2 + \frac{m_\pi}{2m_n} \right]$$

Here, m_n, m_π are the nucleon and pion masses.

39. For nucleons and pions the charge of any member of the multiplet is given by the expression

$$Q = I_3 + \frac{1}{2}B$$

Here, I_3 = third component of the isotopic spin.

40. The Gell-Mann-Okulo mass formula is

$$m = m_0 + m \cdot Y + m_2 \left(I(I+1) - \frac{1}{4}Y^2 \right)$$

$$Y = B + S = 2(I - 1)$$

Here, I = spin.

MULTIPLE CHOICE QUESTIONS

1. A negative nuclear quadrupole moment indicates that nucleus is
   (a) prolate   (b) oblate   (c) spherical (d) spheroidal

2. The cross section for emission of bremsstrahlung depends on the target atomic number Z as
   (a) Z^{-1} (b) Z^{-2} (c) Z (d) Z^2

3. In a linear accelerator the length L^n of the nth drift tube is proportional to
   (a) $\sqrt{n}$ (b) $3\sqrt{3}$ (c) n^2 (d) n

4. There is a gap in the following series of magic numbers. Identify the number in the gap 2, 8, 20, 28, ..., 82, 126
   (a) 71   (b) 60   (c) 59   (d) 50

5. Which of the following has the zero spin
   (a) neutron   (b) neutrino
   (c) μ-mesons   (d) π-mesons

6. The nuclear forces have at least in part, an exchange character and they are like
   (a) molecular forces
   (b) atomic forces
   (c) ionic forces
   (d) van der Waals forces

7. The nuclear force is
   (a) charge-dependent
   (b) charge-independent
   (c) spin-independent
   (d) weakly in triplet state

8. The nuclei having mass number as a multiple of four and n/p ratio equal to unity will be
   (a) most unstable
   (b) most stable
   (c) early stable
   (d) non-existent

9. For the magic number nuclei, the quadrupole moments and the neutron capture cross section show remarkably
   (a) low values
   (b) negative values
   (c) very high values
   (d) zero values

10. The mean life time of excited state is neglected with half level widths by
    (a) $\dfrac{\hbar}{\Gamma}$
    (b) $\dfrac{\Gamma}{\hbar}$
    (c) $\hbar\Gamma$
    (d) $\hbar\Gamma^2$

11. Sharp resonances occur when neutron energy is of the order of
    (a) 0.1 MeV
    (b) 10 MeV
    (c) 100 MeV
    (d) 1000 MeV

12. β-decay is not the disintegration of free neutrons but is the overall property of
    (a) neutron
    (b) proton
    (c) atom
    (d) nucleus

13. The internal conversion electron produces
    (a) a continuous spectrum
    (b) a series of polyenergetic lines
    (c) a series of monoenergetic lines
    (d) fluorescence

14. The photons can interact with particles only through their
    (a) frictional force
    (b) magnetic field
    (c) thermal energy
    (d) electric charge

15. The fission rate for U^{235} required to produce 1 watt is
    (a) 3.13×10^{10} fission/sec   (b) 3.13×10^{20} fission/sec
    (c) 2.13×10^{10} fission/sec   (d) 31.3×10^{10} fission/sec

16. The magnetic moment of deutron is due to the
    (a) addition of proton and neutron magnetic moments
    (b) difference between proton and neutron magnetic moments
    (c) product of proton and neutron magnetic moments
    (d) square of the proton magnetic moment

17. The neutron–neutron scattering is not practically possible because
    (a) the neutrons are chargeless particles
    (b) of the non-availability of neutron target
    (c) of the availability of neutron target
    (d) of the p–p forces

18. For the nuclei first beyond the closed shells, quadrupole moment is
    (a) positive   (b) zero
    (c) negative   (d) very high value

19. The collective model predicts about the rotational levels of
    (a) even–odd nuclei   (b) odd–odd nuclei
    (c) even–even nuclei   (d) even–odd–even nuclei

20. In nuclear reactions the magnetic dipole moment of reacting nuclei is
    (a) zero   (b) conserved
    (c) not conserved   (d) partly conserved

21. The instrument most often used for measuring charged particle energies are
    (a) cloud chamber   (b) Geiger–Müller counter
    (c) scintillation counter   (d) quadract electrometer

22. In β-decay
    (a) conservation of energy holds good
    (b) conservation of linear momentum holds good
    (c) conservation of angular momentum does not violate
    (d) conservation of energy does not hold good

23. The angular momentum of neutrino is
   (a) $\hbar$ (b) $\dfrac{\hbar}{2}$ (c) $\dfrac{\hbar}{3}$ (d) $\dfrac{\hbar}{4}$

24. Nuclides with mass number 230 or more exhibit
   (a) induced fission
   (b) photo fission
   (c) spontaneous fission
   (d) stimulated fission

25. Gravitions have a spin of
   (a) $\hbar$ (b) $2\hbar$ (c) $3\hbar$ (d) $4\hbar$

26. Given that the binding energy of 2He, 4He and 6Li are B_2, B_4 and B_6 respectively, the Q of the reaction $2H + 6Li \rightarrow 2\,^4He + Q$ is given by
   (a) $2B_2 + 6B_6 - 8B_4$
   (b) $B_2 + B_6 - 2B_4$
   (c) $2B_2 - B_2 - B_6$
   (d) $8B_4 - 6B_6 - 2B_2$

27. A new quark the top quark t has been recently discovered. It carries a Baryon number $= \dfrac{1}{3}$ and electric charge $\dfrac{2}{3}t$ is its antiparticle. Which of the following sets of quarks can combine to yield a baryon or a meson?
   (a) tu (b) $\bar{t}du$ (c) tdd (d) $td\bar{d}$

28. The probability T of α-decay from a nucleus at rest is given in terms of the radius of nucleus R and the kinetic energy K of the α-particle by one of the following formulae. Use dimensional analysis to determine the correct one. [C_1 and C_2 are numbers. You may recall $\dfrac{e^2}{\hbar \varepsilon_0 C}$ is dimensionless]

   (a) $\ln T = C_1 \dfrac{e}{\hbar} \left(\dfrac{mR}{\varepsilon_0} \right)^{\frac{1}{2}} + C_2 \left(\dfrac{e^2}{\hbar \varepsilon_0} \right) \left(\dfrac{m}{K} \right)^{\frac{1}{2}}$

   (b) $\ln T = C_1 \dfrac{e}{\hbar} \left(\dfrac{m}{\varepsilon_0 R} \right)^{\frac{1}{2}} + C_2 \left(\dfrac{e^2}{\hbar \varepsilon_0} \right) (mK)^{\frac{1}{2}}$

   (c) $\ln T = C_2 \left(\dfrac{e^2}{\hbar \varepsilon_0} \right) \left(\dfrac{mR}{K} \right)^{\frac{1}{2}}$

   (d) $\ln T = C_1 \dfrac{e}{\hbar} \left(\dfrac{mR}{\varepsilon_0 K} \right)^{\frac{1}{2}}$

29. According to nuclear shell model which includes spin orbit coupling, the spin and parity of the ground state of 5^4B is

    (a) $\left(\dfrac{3}{2}\right)^-$ (b) $\left(\dfrac{3}{2}\right)^+$ (c) $\left(\dfrac{1}{2}\right)^+$ (d) $\left(\dfrac{1}{2}\right)^-$

30. Nuclei which are β^- emitters lie
    (a) below the lines of β stability
    (b) on the line of β stability
    (c) above the line of β stability
    (d) below $N = Z$ line

31. The binding energy per nucleons is maximum for the nucleus
    (a) ^{56}Fe (b) ^4He (c) ^{208}Pb (d) ^{101}Mo

32. The quark structure of Δ^{++} is
    (a) UUU (b) Udu (c) SSS (d) ddd

33. Given that the binding energy of hydrogen atom in the ground state is 13.6 eV, the binding of (hydrogen atom) energy for $n = 2$ state of the positronium atom is
    (a) 13.6 eV   (b) 6.8 eV   (c) 3.4 eV   (d) 1.7 eV

34. The angular momentum and parity of $_8O^{17}$ nucleus according to the nuclear shell model (including spin orbit) is

    (a) 0^+ (b) $\left(\dfrac{1}{2}\right)^-$ (c) $\left(\dfrac{3}{2}\right)^+$ (d) $\left(\dfrac{5}{2}\right)^+$

35. Nuclear force is
    (a) spin-dependent
    (b) spin-independent
    (c) dependent only on total isospin
    (d) dependent only on the third component of isospin

36. The ground state of deuteron is
    (a) a pure s-state   (b) a pure d-state
    (c) a pure p-state   (d) a mixture of s and d states

37. Even nuclei have total ground state angular momentum J equal to
    (a) the vector sum of the unpaired neutron and unpaired proton j values
    (b) the vector sum of the last neutron and the last proton j values

(c) zero
(d) unit always

38. In gamma decay the relation between mean life of the source T and the line width Γ (at half amplitude) is

    (a) $\Gamma = \Gamma$ (b) $\Gamma = \dfrac{T}{\hbar}$ (c) $\Gamma = \dfrac{\hbar}{T}$ (d) $\Gamma = \hbar T$

39. In the Gamow–Teller transitions, the interactions involved are
    (a) scalar and vector
    (b) vector and tensor
    (c) tensor and axial vector
    (d) tensor and pseudo scalar

40. An antineutrino has
    (a) positive helicity and zero spin and linear momentum
    (b) negative helicity and its spin and linear momentum are in opposite directions
    (c) negative helicity and its spin and linear momentum are in the same directions
    (d) positive helicity and its spin is along that of its linear momentum

41. The four-factor formula $k = \eta \varepsilon p f$ will represent in criticality condition when
    (a) $\eta = 1$ (b) $p = 1$ (c) $k = 1$ (d) $\varepsilon = 1$

42. When the multiplication factor $K > 1$ then the reactor is said to be
    (a) subcritical
    (b) critical
    (c) supercritical
    (d) out of order

43. Proton and neutron are classified as a
    (a) non-strange baryons
    (b) strangeness-1-baryon
    (c) strangeness-2-baryons
    (d) non-strange meson

44. Which of the following reactions is forbidden by baryon conservation?
    (a) $k^+ \to \pi^\circ e + \gamma_e$
    (b) $k^- + d \to \pi^+ + \Sigma^-$
    (c) $k^+ \to \overset{\circ}{\Lambda} \mu + \gamma_\mu$
    (d) $\mu^- + p \to \overset{\circ}{\Lambda} + \gamma_\mu$

45. Nuclear force is
    (a) tensor force only
    (b) long range force only
    (c) central force only
    (d) none of the above

46. Pick the isospin triplet.
    (a) π-mesons (b) nucleons
    (c) μ-mesons (d) quarks

47. In the shell model the spin and parity of the nucleons is determined by the spin and parity of the
    (a) nucleons(s) unpaired (b) neutrons(s) unpaired
    (c) proton(s) unpaired   (d) pions inside

48. Non-conservation of parity is observed in
    (a) alpha decay only (b) β-decay only
    (c) gamma decay only (d) pion decay

49. In an electric dipole transition, the change in angular momentum is
    (a) 0 (b) $\frac{1}{2}$ (c) 1 (d) 2

50. In Fermi transition the interactions involved are
    (a) scalar
    (b) vector
    (c) tensor and axial vector
    (d) scalar and pseudo scalar

51. The maximum energy lost by a neutron in a single elastic collision
    (a) decreases with decreasing mass of the target nucleus
    (b) increases with increasing mass of the target nucleus
    (c) decreases with increasing mass of the target nucleus
    (d) is 1 MeV

52. The geometrical buckling factor for a base spherical reactor is (R-radius)
    (a) πR^2 (b) $\left(\frac{\pi}{R}\right)^2$ (c) $\frac{4}{3}\pi R^3$ (d) $4\pi R^2$

53. $\pi^{\pm}, \pi^0$ particles are classified as
    (a) leptons
    (b) non-strange mesons
    (c) strange mesons
    (d) non-strange baryons

54. Which of the following reactions is forbidden

   (a) $\mu + p \to \overset{\circ}{\Lambda} + \gamma_\mu$
   (b) $K^+ \to \overset{\circ}{\pi}_e + \gamma_e$
   (c) $n \to pe^- \bar{\gamma}_e$
   (d) $\Sigma^+ \to \overset{\circ}{\lambda}e^+ \gamma_e$

55. In an exothermal nuclear reaction,
   (a) nuclear energy is converted into mass
   (b) nuclear mass is converted into energy
   (c) total nuclear rest mass is conserved
   (d) total nuclear KE is conserved

56. The first excited state of C^{12} is at 4.43 MeV. Inelastic scattering between neutron and C^{12} cannot occur unless the neutron energy in the lab frame is at least
   (a) 4.1 MeV (b) 4.43 MeV (c) 4.6 MeV (d) 4.8 MeV

57. Fermi age τ is given by $\tau =$

   (a) $\left(\frac{1}{6}r^{-2}\right)$
   (b) $\left(\frac{1}{6}\right)r^{-2}$
   (c) $\left(\frac{1}{6}r^{-2}\right)^{\frac{1}{2}}$
   (d) $\left(\frac{1}{6}r^{-2}\right)^{\frac{1}{2}}$

58. In the Fermi four-factor formula $k\alpha = \varepsilon t p \eta$, which two factors are greater than unity?
   (a) t and η (b) ε and t (c) p and f (d) f and η

59. Figure shows the neutron fluxes for the region bare thermal reactions which are in the shape in infinite slab (Is) infinite cylinder (Ic) and sphere (sp). Identify the graphs with the geometry.

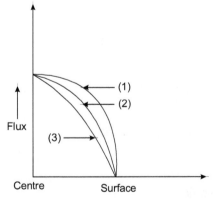

   (a) 1 is Is, 2 is Ic, 3 is sp
   (b) 2 is Is, 3 is Ic, 1 is sp
   (c) 3 is Is, 1 is Ic, 2 is sp
   (d) 2 is Is, 1 is sp, 3 is Ic

60. For which of the following bare critical geometries is the leakage of neutrons per volume of fissionable material the least?
    (a) parallelopiped   (b) cylinder
    (c) sphere   (d) cone

61. An important fission product poison is
    (a) Kr^{86} (b) Xe^{135} (c) Ba^{135} (d) Fe^{56}

62. Which of the following statements is true?
    (a) delayed neutrons decrease the reactor period
    (b) delayed neutrons increase the reactor period
    (c) delayed neutrons do not affect the reactor period
    (d) the total yield of delayed neutrons per fission is 0.2

63. Which of the following materials can be used for making control rods in thermal reactors?
    (a) Be   (b) Ag   (c) Pb   (d) Hofnium

64. Derivation of Fick's law assumes
    (a) neutron scattering is not isotropic in the laboratory frame
    (b) neutron flux is a function of time
    (c) neutron flux is a slowly varying function of position
    (d) neutron source distribution in the medium is Gaussian

65. The s-process stops with heavy element
    (a) ^{209}Pb (b) ^{209}Bi (c) ^{210}Po (d) ^{235}U

66. The burning of heavy nuclei is
    (a) more sensitive to temperature
    (b) less sensitive to temperature
    (c) independent of temperature
    (d) not possible at all

67. A zero mass quark, moving in a cavity of radius R exerts a pressure on the wall of the cavity which is proportional to
    (a) $\frac{1}{R^4}$ (b) $\frac{1}{R^2}$ (c) R^2 (d) R^4

68. Consider a virtual process $p \to n\pi^+$, $n \to p\pi^-$ and $p \to p\pi°$. The amplitudes for the three are in a ratio
    (a) $1:1:\frac{1}{\sqrt{2}}$ (b) $1:1:1$ (b) $1:1:0$ (d) $1:0:1$

69. A Lagrangian with a doublet of Dirac fields and triplet of solar fields is invariant under isospin su (2) transformations. The number of conserved quantities is
    (a) 5   (b) 3   (c) 2   (d) 1

70. Which of the following is (Abelian) guage invariant? A^μ is the guage field
    (a) $\bar{\psi}\gamma_\mu \psi A^\mu$
    (b) $\bar{\psi}\sigma_{\mu\nu} \psi F^{\mu\nu}$
    (c) $\bar{\psi}\psi A_\mu A^\mu$
    (d) $\bar{\psi}\partial_\mu \psi A^\mu$

71. Q_{CD} is asymptotically free when the number of flavours is
    (a) <6 (b) <4 (c) >2 (d) ≤ 16

72. The perturbative approach to Q_{CD} breaks down when the energies are of the order of
    (a) 1000 GeV   (b) 100 GeV
    (c) 10 GeV   (d) 300 MeV

73. If the bag constant $B = 60$ MeV/fm^3, then at the critical temperature zero, Baryon chemical potential is nearly
    (a) 50 MeV   (b) 100 MeV   (c) 937 MeV   (d) 1 GeV

74. The energy E and the thermodynamics potential Ω of a quark is
    (a) $3E + 4\Omega < 0$ (b) $3E - 4\Omega = 0$
    (c) $3E - 4\Omega > 0$ (d) $E + 3\Omega > 0$

75. The nuclear force is
    (a) always attractive   (b) always repulsive
    (c) always central   (d) none of the above

76. If the magnetic moment of proton is 2.79 nm, then the magnetic moment of deuteron will be
    (a) zero   (b) > 2.79 nm
    (c) 2.79 nm   (d) < 2.79 nm

77. In the Schmidt model, the magnetic moment is
    (a) entirely due to the odd neutron
    (b) entirely due to the odd nucleon
    (c) entirely due to the odd proton
    (d) due to all nucleons present

78. In compond nucleus reaction, the intermediate nucleus is normally
    (a) in its ground state
    (b) highly stable
    (c) the same as the final nucleus
    (d) in its excited state

79. Tensor and axial vector interactions appear in the allowed
    (a) Fermi transitions
    (b) Gamow–Teller transition
    (c) dipole transition
    (d) monopole transition

80. For E-multipole radiations
    (a) $\pi_i = \pi_f$
    (b) $\pi_i = (-1)^l \pi_f$
    (c) $\pi_i = -\pi_i$
    (d) $\pi_i = (-1)^l \pi_f$

81. In the Fick's law expression for neutron, $J = \text{div grad}(\phi)D$ has the dimension of
    (a) mass
    (b) length
    (c) time
    (d) charge

82. Identify the material used for fabricating control rods in thermal reactors
    (a) copper
    (b) zinc
    (c) cadmium
    (d) carbon

83. π^+, π^0 particles are classified as
    (a) leptons
    (b) non-strange mesons
    (c) strange mesons
    (d) non-strange baryons

84. What is X in $n \to p + \bar{e} + X$?
    (a) e^+
    (b) $\dfrac{\gamma}{\mu}$
    (c) $\dfrac{\gamma}{e}$
    (d) $\dfrac{\bar{\gamma}}{e}$

85. The ratio of the sizes of $_{82}Pb^{208}$ and $_{12}Mg^{26}$ nuclei is approximately
    (a) 2
    (b) 4
    (c) 8
    (d) 16

86. Which of the following statements is incorrect for the nuclear force between two nucleons
    (a) it is charge-independent
    (b) it is spin-independent
    (c) it is velocity-independent
    (d) it has a noncentral component

87. The maximum energy of deuteron coming out of a cyclotron accelerator is 20 MeV. The maximum energy of proton that can be obtained from this accelerator is
    (a) 10 MeV   (b) 20 MeV
    (c) 30 MeV   (d) 40 MeV

88. For the deduction of neutrons one should utilize
    (a) high purity detector
    (b) scintillation detector
    (c) silicon surface barrier detector
    (d) proportional counter field with BF_3

89. Positronium is a hydrogen like bound state of positron and an electron. If the nth energy level of hydrogen atom is given by $E_n^4 = \frac{-1}{(4\pi\varepsilon_0)^2} \frac{m_0^4}{2n^2\hbar^2}$ the nth energy level E_n^p of the positronium will be equal to
    (a) $\frac{1}{2}E_n^4$ (b) E_n^4 (c) $\frac{3}{2}E_n^4$ (d) $2E_n^4$

90. The bound charge density in a sphere of radius R carrying polarization $\vec{P} = K\vec{r}$ is
    (a) K (b) $\frac{K}{r}$ (c) $-3K$ (d) zero

91. At large distance the electric field due to a quadrupole varies as
    (a) $\sim \frac{1}{r^3}$ (b) $\sim \frac{1}{r^6}$ (c) $\sim \frac{1}{r^4}$ (d) $\frac{1}{r^5}$

92. The $_{92}U^{238}$ emits 8 α-particles and n β-particles before it becomes stable nucleus of lead of atomic number 82 then n is
    (a) 8   (b) 6   (c) 4   (d) 2

93. The nuclear process in which the law of conservation of parity is violated is
    (a) α-decay (b) β-decay (c) γ-decay (d) fission

94. Which nuclear model explains nuclear fission
    (a) shell model
    (b) collective model
    (c) liquid drop model
    (d) optical model

95. Electric quadrupole is zero if it is in the
    (a) s-state
    (b) p-state
    (c) d-state
    (d) f-state

96. The elementary particles for which the spin is zero is
    (a) neutron
    (b) π-meson
    (c) k-meson
    (d) μ-meson

97. The instrument which is suitable for absolute measurement of the activity of a β-active source is
    (a) GM counter
    (b) scintillation counter
    (c) proportional counter
    (d) ionization counter

98. Neutrons can be detected by
    (a) proportional counter
    (b) ionization counter
    (c) GM counter
    (d) scintillation counter

99. Which of the following materials is used in the detection of neutrons?
    (a) Si(Li) (b) NAI(TL) (c) BF_3 (d) Ge(Li)

100. For the self sustained fission reaction the multiplication factor should be
    (a) $>>1$ (b) <1 (c) $=1$ (d) $<<1$

101. Which of the following particles cannot be accelerated in a cyclotron?
    (a) electron
    (b) proton
    (c) deutron
    (d) alpha particles

102. If λ is the disintegration constant for α emission and R the range of α-particles emitted, the two are related by the equation
    (a) $\lambda = AR + B$
    (b) $\lambda = AR^2 + B$
    (c) $\lambda = Ae^R + B$
    (d) $\log \lambda = \log R + B$

103. The absorption of gamma rays by the process of pair production is prominent at
    (a) low Z, low E
    (b) low Z, high E
    (c) high Z, low E
    (d) high Z, high E

104. In the sequence 8, 20, 50, 126 of magic numbers, which number is missing?
    (a) 80   (b) 52   (c) 84   (d) 86

105. In the semiempirical mass formula the contribution of the pairing term is proportional to
    (a) $A^{\frac{1}{4}}$ (b) $A^{-\frac{1}{4}}$ (c) $A^{-\frac{1}{2}}$ (d) $A^{-\frac{3}{4}}$

106. The radius r of a nucleus of mass number A, depends on A as
    (a) $r \propto A$
    (b) $r \propto A^2$
    (c) $r \propto A^{\frac{1}{3}}$
    (d) r is independent of A

107. Maintaining of chain fission reaction requires that the sum of a fissile material exceeds a critical mass. The reason is
    (a) the system must have a minimum total energy
    (b) the surface-to-volume ratio of the system must be small so that the neutron production exceeds the loss
    (c) the size of the system must exceed neutron wavelength
    (d) none of the above

108. Is parity conserved in (i) strong interactions (ii) weak interactions
    (a) (i) yes (ii) yes
    (b) (i) yes (ii) no
    (c) (i) no (ii) yes
    (d) (i) no (ii) no

109. The working of a cyclotron depends critically on the fact that the angular frequencies of charged particles in a magnetic field
    (a) does not depend on the radius of its orbit
    (b) depends linearly on its radius
    (c) does not depend on its mass
    (d) depends on the electric field between the dees

110. Which of the following characterizes the π-mesons?
    (a) baryon and boson
    (b) baryon and fermion
    (c) lepton and boson
    (d) lepton and fermion

111. Neutrons are scattered by phonons of a solid because
    (a) neutron size is smaller than lattice spacing
    (b) neutrons can interact with electrons through weak interaction
    (c) neutrons can interact with ions of the lattice
    (d) neutron energy is comparable to binding energy of the atom in the lattice

112. In the semi-empirical binding energy formula, the term which varies as $A^{-\frac{1}{3}}$ (A being the atomic mass number) is due to
    (a) surface tension   (b) volume energy
    (c) Coulomb energy   (d) symmetry effect

113. Which of the following statements correctly describes the properties of σ neutrino?
    (a) zero charge, zero spin, F-D statistics
    (b) zero charge, half spin, F-D statistics
    (c) zero charge, half spin, BE statistics
    (d) zero charge, zero spin, BE statistics

114. Which of the following is preferred for acceleration of electrons?
    (a) betatron   (b) cyclotron
    (c) Van de Graff generator   (d) linear accelerator

115. The Compton cross section for absorption of γ-rays by matter depends on the atomic number z as
    (a) z (b) z^{2-3} (c) z^3 (d) z^5

116. The electron is a particle in category of
    (a) hyperons   (b) leptons
    (c) mesons   (d) baryons

117. The absorption of γ-rays by matter by photoelectric effect is most prominent with
    (a) low-energy photons and light elements
    (b) low-energy photons and heavy elements
    (c) high-energy photons and light elements
    (d) high-energy photons and heavy elements

118. If a nucleus having spin I and quadrupole moment Q is non-spherical, it necessarily means
    (a) $Q \neq 0$ (b) $I \neq 0$ (c) $Q = 0$ (d) $I = 0$

119. The semi-empirical mass formula, the dependence of the pairing energy on the mass number A is given by

(a) $A^{\frac{2}{3}}$ (b) $A^{-\frac{1}{3}}$ (c) A (d) $A^{-\frac{3}{4}}$

120. The stopping power of a heavy charged particle passing through matter depends on its
    (a) velocity only
    (b) charge only
    (c) mass only
    (d) velocity and charge

121. The equation that connects the charge, the isospin vector, the baryon number and the strangeness is called?
    (a) Klein-Nishina formula
    (b) Breit-Weigner formula
    (c) Klein-Gordon equation
    (d) Gell-Mann-Nishijima formula

122. For a fission reaction to be steady, the effective multiplication factor is
    (a) 1 (b) < 1 (c) > 1 (d) 0

123. The NaI (TI) detector, the radian is due to transition between
    (a) NaI conduction and valence bonds
    (b) Activator levels
    (c) NaI conduction band and activator level
    (d) Activator level and NaI valence bands

124. The Boron isotope which is useful in neutron detection has mass number
    (a) 8 (b) 9 (c) 10 (d) 11

125. Which of the following statements concerning deuteron is true?
    (a) The binding energy per nucleon is more than 25 MeV and hence it is tightly bound to two nucleon systems.
    (b) Its magnetic moment is less than the sum of the magnetic moments of its constitutents.
    (c) Its ground state spin is the same as that of the positron and hence is a boson.
    (d) The existence of deuteron implies that the nuclear forces are central and have a range of at least 100 fm.

126. The gamma decay $2^+ \to 0^+$ is
   (a) a pure E_2 transition
   (b) a pure M_2 transition
   (c) a mixture of E_2 and M_2 transition
   (d) a mixture of E_1 and M_3 transitions

127. According to the shell model of the nucleus, the ground state spin, parity and magnetic moment of $3H$ are respectively
   (a) $\frac{1}{2}, +, 1.91$ nm
   (b) $\frac{1}{2}, +, 2.79$ nm
   (c) $\frac{1}{2}, -, 1.91$ nm
   (d) $\frac{1}{2}, -, 2.79$ nm

128. Which of the following statements concerning nuclear reactions is false?
   (a) There can be reaction without scattering.
   (b) There cannot be reaction without scattering.
   (c) There can be scattering without any reaction.
   (d) The phenomenon of resonance has been observed in nuclear reaction.

129. A neutron is
   (a) a hadron, but not fermion
   (b) a lepton, but not a baryon
   (c) a boson, but not a meson
   (d) None of the above

130. Choose the correct answer regarding a synchrotron
   (a) It can accelerate neutrons.
   (b) It can detect neutrons.
   (c) It was designed to take into account the variation of the mass of a proton with speed.
   (d) Its design is due to the variation of the spin of the proton with velocity.

131. Two radioactive samples X (half-life 3 years) and Y (half-life 2 years) have been decaying for many years. Today the number of atoms in the sample X is twice the number of atoms in the sample Y. Both the samples had the same number of atoms
   (a) 4 months ago
   (b) 3 years ago
   (c) 8 years ago
   (d) never before

132. The threshold gamma ray energy for electron–positron pair production is
   (a) 0.102 MeV   (b) 1.02 MeV
   (c) 10.2 MeV   (d) 102 MeV

133. The cyclotron frequency for non-relativistic particle of mass M and charge q in magnetic field of induction B is given by
   (a) $-\omega = \dfrac{M}{q}(B)$ (b) $\omega = qBM$
   (c) $\omega = -BMq$ (d) $\omega = \dfrac{qB}{M}$

134. The reaction $n + p \to d + \gamma$ occur primarily via
   (a) strong and weak interactions
   (b) weak and electromagnetic interactions
   (c) strong and electromagnetic interactions
   (d) purely electromagnetic interaction

135. The nuclear force is
   (a) repulsive in nature   (b) charge-independent
   (c) a long range force   (d) a very weak force

136. The selection rule for Fermi transition is
   (a) $\Delta I = 0$ (b) $\Delta I = 1$
   (c) $\Delta I = -1$ (d) $\Delta I = \pm 1$
   where, I is the nuclear spin.

137. The strangeness quantum number associated with a K⁺ meson is
   (a) 0   (b) –2   (c) –3   (d) 1

138. The decay chain for the $_{92}U^{238}$ nucleus involves 8 α-decays and 6 β-decays. The final nucleus at the end of the process will have
   (a) $Z = 82, A = 206$ (b) $Z = 84, A = 224$
   (c) $Z = 88, A = 206$ (d) $Z = 76, A = 200$

139. The possible electromagnetic multiple transition between the nuclear state
$$I^p = \left(\frac{3}{2}\right)^+ \text{ and } I^p = \left(\frac{5}{2}\right)^+ \text{ are}$$

(a) M_1, M_2　　　　　　　(b) M_1, E_1
(c) M_2, E_3　　　　　　　(d) M_1, E_2, M_3, E_4

140. The deviation of the charge distribution of a nucleus from spherical symmetry can be estimated by measuring its
    (a) electric charge
    (b) electric dipole moment
    (c) magnetic dipole moment
    (d) electric quadrupole moment

141. The ground state of a deuteron is a
    (a) a pure g_{S1} state
    (b) pure g_{P1} state
    (c) mixture of g_{S1} and g_{P1} states
    (d) mixture of g_{S1} and g_{d1} states

142. The contribution to the total binding energy of the nucleus $_2X^4$ by the surface term is proportional to
    (a) $A^{-\frac{2}{3}}$　　(b) $A^{-\frac{1}{3}}$　　(c) $A^{\frac{1}{3}}$　　(d) $A^{\frac{2}{3}}$

143. When a neutron and proton combine to form a deuteron, the amount of energy given off is
    (a) 20.22×10^6 eV　　　　(b) 200 eV
    (c) 2.22×10^6 eV　　　　(d) 22 eV

144. The thermal neutrons have energy equal to
    (a) 0.25 eV　(b) 0.025 eV　(c) 2.5 eV　(d) 0.0025 eV

145. The elementary particle that never decays in the free space is
    (a) μ- meson　　　　　　(b) hyperons
    (c) proton　　　　　　　　(d) neutron

146. The radiation detector that does not depend on the ionization of the gas is
    (a) Geiger counter　　　　(b) photographic emulsion
    (c) electroscope　　　　　(d) cloud chamber

147. The quadrupole moment of nucleus is positive if its square is
    (a) oblate　　　　　　　　(b) prolate
    (c) spherical　　　　　　(d) linear

148. Bartlett exchange force arises from
    (a) spatial exchange
    (b) spatial and spin exchange
    (c) charge exchange
    (d) spin exchange

149. A reactor is said to be critical when
    (a) a chain reaction becomes self sustained
    (b) the reaction is shut down
    (c) the reactor becomes radioactive and causes health hazards
    (d) reactor builds

150. The life time of the compound nucleus is approximately
    (a) 10^{-12} sec
    (b) 1 sec
    (c) 10^{-21} sec
    (d) 1 m sec

151. The radiation due to accelerated electron is called
    (a) Cerenkov radiation
    (b) Bremstrahlung radiation
    (c) black body radiation
    (d) gamma radiation

152. The binding energy of a nucleon in a nucleus of large mass number is
    (a) 8 eV (b) 8 MeV (c) 931 MeV (d) 931 eV

153. The liquid drop model of nucleus is based on
    (a) weak interaction of the nucleons
    (b) strong interaction among the nucleons
    (c) electromagnetic interaction among the nucleons
    (d) parity

154. The stripping reactions are nuclear reactions involving target nuclei and
    (a) protons
    (b) deuterons
    (c) neutrons
    (d) X-rays

155. U^{235} has a large cross section for fission by slow neutrons than U^{238}, the reason being
    (a) U^{238} has a large mass number
    (b) quantum mechanical pairing energy
    (c) U^{235} has a small mass number
    (d) neutron has a magnetic moment

156. The reason for deuteron processing negligible electric dipole moment is
    (a) parity is not a good quantum number
    (b) parity is not concerned in weak interaction
    (c) parity is a good quantum number
    (d) none of the above

157. The nuclear forces include
    (a) central force only
    (b) tensor forces only
    (c) combination of the above two forces
    (d) long range forces

158. Schmidt lines are related to the
    (a) magnetic momentum of the nucleus
    (b) quadrapole momentum of the nucleus
    (c) size of the nucleus
    (d) charge of the nucleus

159. The ground state of deuteron is
    (a) pure s-state
    (b) pure p-state
    (c) pure d-state
    (d) mixture of d and s-states

160. Identify the compound nucleus for the two reactions $O^{16}(d, p)O^{17}$ and $N^{14}(\alpha, p) O^{17}$?
    (a) $\overset{*}{F}{}^{16}$
    (b) $\overset{*}{F}{}^{17}$
    (c) $\overset{*}{F}{}^{18}$
    (d) $\overset{*}{O}{}^{17}$

161. In Fermi-transition (β-decay) the interaction involved are
    (a) scalar
    (b) scalar and vector
    (c) tensor and axial vector
    (d) scalar and pseudo scalar

162. For M-multipole radiation
    (a) $\pi_i = \pi_f$
    (b) $\pi_i = (-)^l \pi_f$
    (c) $\pi_i = -\pi_f$
    (d) $\pi_i = -(-1)^l \pi_f$

163. The ratio of the number of neutrons produced by fission in any one generation to the number in the immediately preceding generation is called
    (a) multiplication factor
    (b) thermal utilization factor
    (c) fast fission factor
    (d) fission fraction

164. The unit for Fermi age is
    (a) cm² (b) cm (c) s (d) s²

165. Which of the following reaction is forbidden
    (a) $\mu^- \to p \to \overset{\circ}{\Lambda} + \gamma_\mu$
    (b) $k^+ \to \pi^0 + e^+ + \upsilon_e$
    (c) $n \to p + e^- + \bar{\upsilon}_e$
    (d) $\Sigma^+ \to \pi^0 + e^+ + \upsilon_e$

166. $\mu^- + p \to X + Y$ where X and Y are
    (a) n, γ_μ (b) n, e^+ (c) n, e^- (d) $n, \bar{\gamma}_\mu$

167. Isospin triplet is
    (a) triton
    (b) π-mesons
    (c) μ-mesons
    (d) quarks

168. Which of the following reactions is forbidden by energy conservation?
    (a) $\Sigma^+ \to \Lambda^0 + \mu^+ + \upsilon_\mu$
    (b) $k^+ + d \to \pi^+ + \Sigma^-$
    (c) $k^+ \to \pi^0 + e^+ + \upsilon_e$
    (d) $\mu^- + P \to \Lambda^0 + \upsilon_\mu$

169. The charge distribution in the ground state of deuteron is
    (a) spherical shape
    (b) oblate shaped
    (c) prolate shape
    (d) none of the above

170. Identify the compound nucleus for the reaction $_5B^{11}(p - \alpha)$
    (a) $(_5B^{10})^*$ (b) $(_5C^{12})^*$ (c) $(_4B^{10})^*$ (d) $(_5C^{11})^*$

171. Non-conservation of parity is observed in
    (a) alpha decay
    (b) beta decay
    (c) gamma decay
    (d) muon decay

172. Tensor and axial vector interactions are involved in the allowed
    (a) dipole transitions
    (b) Gamow-Teller transitions
    (c) monopole transition
    (d) Fermi transition

173. The maximum energy lost by a neutron in a single elastic collision
    (a) is 15.1 MeV
    (b) is 0.5 eV
    (c) increases with increasing mass of the target nucleus
    (d) decreases with decreasing mass of the target nucleus

174. When the multiplication factor $k > 1$ then the reactor is said to be
    (a) of control
    (b) critical
    (c) subcritical
    (d) supercritical

175. π^+ and π^0 are particles classified as
    (a) non-strange baryons
    (b) leptons
    (c) non-strange mesons
    (d) strange mesons

176. Which of the following reactions is forbidden by baryon number conservation
    (a) $k^+ \to \pi^0 + e^+ + v_e$
    (b) $\Sigma^+ \to \Lambda^0 + \mu^+ + v_\mu$
    (c) $k^- + d \to \pi^+ + \Sigma^-$
    (d) $\mu^- + P \to \Lambda^0 + v_\mu$

177. In the optical model of the nucleus
    (a) the nuclear potential is complex
    (b) the nuclear refractive index is complex
    (c) the nucleus is transparent
    (d) scattering cannot be treated

178. Scalar and vector interactions are involved in the allowed
    (a) Gamow–Teller transition
    (b) fermi transition
    (c) dipole transition
    (d) multiple transition

179. In the four-factor formula $K = \eta \varepsilon p f$. If $K > 1$ then the reactor is in
    (a) supercritical state
    (b) critical state
    (c) subcritical state
    (d) poisoned state

180. Identify the compound nucleus for the reaction $_8O^{18}(p, n)$
    (a) $_9Fe^{19}$
    (b) $_9Fe^{18}$
    (c) $_8O^{19}$
    (d) $_{10}Ne^{19}$

181. Which of the following reactions is allowed under conservation of charge
    (a) $\pi^+ + n \to \Lambda^0 + K^+$
    (b) $\pi^+ + n \to \pi^- + p$
    (c) $k^- \to \pi^0 e^+ v_e$
    (d) $\mu^- + n \to \Lambda^0$

182. The most stable nuclei are
    (a) odd-odd nuclei
    (b) even-even nuclei
    (c) even-odd nuclei
    (d) odd-even nuclei

183. Exchange of spin coordinates gives rise to
    (a) Majorana exchange
    (b) Heisenberg exchange
    (c) Baslett exchange
    (d) Meson exchange

184. The contribution of surface energy to the toal binding energy of the nucleus is proportional to
    (a) $-A^{\frac{2}{3}}$
    (b) $A^{\frac{2}{3}}$
    (c) $A^{\frac{1}{3}}$
    (d) $A^{-\frac{2}{3}}$

185. Which of the following is not a magic number?
    (a) 2
    (b) 8
    (c) 28
    (d) 50

186. For proper working of a GM counter the PD across its electrodes must correspond to
    (a) continuous discharge region
    (b) thousands of volts
    (c) plateau region of its characteristics
    (d) proportional region of its characteristics

187. In a cyclotron with the applied force B, the frequency of applied RF field is
    (a) directly proportional to B
    (b) inversely proportional to B
    (c) directly proportional to B^2
    (d) inversely proportional to B^2

188. The neutrons emitted in beta decay must have
    (a) spin $-\frac{1}{2}$
    (b) spin $+\frac{1}{2}$
    (c) positive charge
    (d) negative charge

189. The total scattering cross section of S-wave neutrons when the phase shift is δ_0 is
    (a) directly proportional to $\sin^2 \delta_0$
    (b) directly proportional to $\sin \delta_0$
    (c) inversely proportional to $\sin^2 \delta_0$
    (d) inversely proportional to $\sin \delta_0$

190. Which of the following reactions is allowed?
    (a) $\pi^- \rightarrow \mu + \bar{\upsilon}_e$
    (b) $\pi^- \rightarrow \bar{\mu} + \upsilon_\mu$
    (c) $\bar{\upsilon}_\mu + P \rightarrow n + e^+$
    (d) $\pi^+ + n \rightarrow k^+ + k^0$

ANSWERS

1. (b)	2. (d)	3. (a)	4. (d)	5. (d)
6. (a)	7. (b)	8. (b)	9. (a)	10. (a)
11. (a)	12. (d)	13. (c)	14. (d)	15. (a)
16. (a)	17. (b)	18. (c)	19. (c)	20. (b)
21. (c)	22. (d)	23. (b)	24. (c)	25. (b)
26. (b)	27. (c)	28. (a)	29. (a)	30. (c)
31. (a)	32. (a)	33. (c)	34. (d)	35. (a)
36. (d)	37. (c)	38. (c)	39. (c)	40. (d)
41. (c)	42. (c)	43. (b)	44. (c)	45. (c)
46. (a)	47. (a)	48. (b)	49. (c)	50. (a)
51. (c)	52. (b)	53. (b)	54. (d)	55. (c)
56. (c)	57. (b)	58. (d)	59. (d)	60. (c)
61. (a)	62. (b)	63. (c)	64. (c)	65. (b)
66. (c)	67. (d)	68. (a)	69. (b)	70. (c)
71. (c)	72. (a)	73. (a)	74. (a)	75. (c)
76. (d)	77. (d)	78. (d)	79. (b)	80. (b)
81. (b)	82. (c)	83. (b)	84. (d)	85. (c)
86. (b)	87. (a)	88. (b)	89. (d)	90. (b)
91. (c)	92. (b)	93. (b)	94. (c)	95. (c)
96. (b)	97. (d)	98. (a)	99. (c)	100. (c)
101. (a)	102. (a)	103. (d)	104. (c)	105. (d)
106. (c)	107. (b)	108. (b)	109. (b)	110. (a)
111. (c)	112. (a)	113. (b)	114. (a)	115. (a)
116. (b)	117. (b)	118. (a)	119. (d)	120. (a)
121. (d)	122. (a)	123. (a)	124. (c)	125. (b)
126. (a)	127. (c)	128. (c)	129. (d)	130. (c)
131. (c)	132. (b)	133. (d)	134. (c)	135. (b)
136. (d)	137. (d)	138. (a)	139. (b)	140. (d)
141. (d)	142. (d)	143. (c)	144. (b)	145. (d)
146. (b)	147. (b)	148. (d)	149. (a)	150. (a)
151. (b)	152. (b)	153. (b)	154. (a)	155. (d)
156. (b)	157. (c)	158. (a)	159. (d)	160. (c)
161. (a)	162. (d)	163. (a)	164. (a)	165. (a)

166. (a)	167. (b)	168. (a)	169. (c)	170. (b)
171. (b)	172. (c)	173. (c)	174. (d)	175. (d)
176. (c)	177. (a)	178. (b)	179. (a)	180. (a)
181. (a)	182. (b)	183. (c)	184. (a)	185. (c)
186. (c)	187. (a)	188. (b)	189. (c)	190. (b)

TEST PAPER 1

1. Force is a vector quantity. For the force to be conservative
   (a) its gradient should vanish
   (b) its curl should vanish
   (c) its divergence should vanish
   (d) gradient, curl and divergence should not vanish

2. A man walks on the road in a straight line with uniform velocity
   (a) his linear momentum is conserved
   (b) his total energy of motion is conserved
   (c) his linear momentum is not conserved
   (d) the total work done is zero

3. The Coriolis force on a particle is related to
   (a) the angular velocity of the rotating frame
   (b) the linear acceleration of the rotating frame
   (c) the linear acceleration of the particle in the rotating frame
   (d) the position vector of the particle in the rotating frame

4. In the generalized coordinates and generalized velocity representations
   (a) all generalized coordinates have the same dimensions
   (b) all generalized velocities have the same dimensions
   (c) all generalized momenta have the same dimensions
   (d) all the above-mentioned respective quantities need not have the same dimensions

5. Translational symmetry implies
   (a) conservation of linear momentum
   (b) conservation of angular momentum
   (c) conservation of total energy
   (d) conservation of the Lagrangian

6. Nutational motion in a symmetric top corresponds to the periodic change in the Eulerian angle (s)
   (a) θ
   (b) ψ
   (c) $\dot{\psi}$
   (d) χ

7. In canonical transformations
   (a) the form of the Lagrangian equations is conserved
   (b) the number of cyclic coordinates can be increased
   (c) the form of Poisson bracket is changed
   (d) generalized velocities become constants of motion

8. Pick the wrong statement.
   (a) The speed of light in free space is independent of the motion of the observer.
   (b) In all inertial frames, the speed of light in free space has the same value.
   (c) Relativistic and Newtonian mechanics agree to relative speeds much lower than the speed of light.
   (d) The speed of light has the same value in all media.

9. Choose the correct answer.
   (a) There is a universal frame of reference.
   (b) The laws of Physics vary from one inertial frame to another.
   (c) In all inertial frames, the laws of Physics remain the same.
   (d) The laws of Physics are different for different observers in relative motion.

10. Which one of the following statements concerning h is correct?
    (a) It has zero dimensions.
    (b) It is dimensionally electron-volt time length
    (c) Its dimensions are the same as those of moment of momentum
    (d) It has the dimensions of force times velocity

11. Given $\psi(x) = 8x^3 - 12x$ and the operator $A = \dfrac{d^2}{dx^2} - 2x \dfrac{d}{dx}$. The eigen value of A is
    (a) 12
    (b) – 6
    (c) – 12
    (d) 6

12. An infinite potential well extends from $x = -a$ to $x = +a$. The wave function of a particle inside this potential
    (a) is only odd
    (b) is only even
    (c) has no restriction
    (d) can either be odd or even

13. In a hydrogen atom, the energy eigen value E_n is related to E, by
    (a) $E_1 + n^2 E_n = 0$
    (b) $E_n + n^2 E_1 = 0$
    (c) $E_n - n^2 E_1 = 0$
    (d) $E_1 - n^2 E_n = 0$

14. Bosons are represented by
    (a) totally antisymmetric wave function
    (b) totally symmetric wave function
    (c) odd wave function
    (d) even wave function

15. According to Fermi's Golden Rule, transition probability per unit time is
    (a) proportional to time
    (b) independent of time
    (c) inversely proportional to density of states
    (d) proportional to the matrix element of the perturbing operator

16. Dirac's relativistic equation has derivatives
    (a) first order in time and first order in space
    (b) first order in time and second order in space
    (c) first order in space and second order in time
    (d) second order in time and second order in space

17. In partial wave analysis
    (a) spherical wave functions are split into plane wave functions
    (b) only high energy scattering can be analysed
    (c) plane wave functions are split into spherical wave functions
    (d) only long range and high potential energy problems can be analysed

18. The potential inside a spherical shell of radius R and charge q is

(a) $\dfrac{1}{4\pi\varepsilon_0} \cdot \dfrac{q}{R^2}$

(b) $\dfrac{1}{4\pi\varepsilon_0} \cdot \dfrac{q}{R}$

(c) $4\pi\varepsilon_0 \cdot \dfrac{R}{q}$

(d) $4\pi\varepsilon_0 R^2 q$

19. Poisson's equation is given by

(a) $\nabla E = -\nabla V$

(b) $\nabla \times E = 0$

(c) $\nabla^2 V = 0$

(d) $\nabla^2 V = -\dfrac{\rho}{\varepsilon_0}$

20. A particle of mass m and charge q executes uniform circular motion of radius R with speed v in a magnetic field B. The condition to sustain this motion is given by

(a) $mv = qBR$

(b) $\dfrac{mv}{R} = \dfrac{q}{B}$

(c) $\dfrac{mv^2}{R} = qB$

(d) $\dfrac{vR}{R} = mqB$

21. If the particle quantum number is 3, then the orbital and magnetic quantum numbers of an electron in the hydrogen atom can be respectively

(a) $\dfrac{3}{2}, \dfrac{3}{2}$ (b) 2, 1   (c) 1, – 2   (d) 3, – 3

22. 10^{23} molecules of hydrogen gas occupy a volume of 20 litres at NTP. The statistics obeyed by this gas
 (a) is Maxwell–Boltzmann statistics
 (b) is Fermi–Dirac statistics
 (c) is Bose–Einstein statistics
 (d) depends on the spin state of the hydrogen

23. The magnetic moment associated with spin of an electron is demonstrated by
 (a) Davisson–Germer experiment
 (b) Franck–Hertz experiment
 (c) Michelson–Morley experiment
 (d) Stern–Gerlach experiment

24. Specific heat at constant pressure for an ideal diatomic gas is
    (a) $C_p = \frac{5}{2}NR$
    (b) $C_p = \frac{1}{2}NR$
    (c) $C_p = \frac{7}{2}NR$
    (d) $C_p = \frac{3}{2}NR$

25. In a grand canonical ensemble
    (a) only the number of particles N is fixed
    (b) only the energy E is fixed
    (c) both E and N are variable
    (d) both E and N are fixed

26. The binding energy per nucleon in a deuteron is
    (a) -522 keV
    (b) 1.111 MeV
    (c) -2.222 MeV
    (d) 2.222 MeV

27. In the liquid drop model of the nucleus of mass A and radius R
    (a) the volume energy is inversely proportional to A
    (b) its surface energy is directly proportional to R
    (c) the coulomb energy is directly proportional to R
    (d) the volume energy makes a positive contribution to the binding energy

28. The enery production in the sun is mainly due to
    (a) fission of carbon nuclei by protons
    (b) proton–proton cycle
    (c) the carbon cycle
    (d) successive capture of neutrons by heavy nuclei

29. Choose the correct statement.
    (a) An S band formed by N atoms can hold N electrons
    (b) A P band formed by N atoms can hold $2N$ electrons
    (c) The $3S$ and $3P$ bands formed by N atoms can hold $6N$ electrons
    (d) The $3S$ and $3P$ bands formed by N atoms can hold $8N$ electrons

30. If the total energy of an electron is four times its rest energy the ratio of its speed to the velocity of light is
    (a) $\sqrt{\frac{15}{4}}$
    (b) $\frac{3}{4}$
    (c) $\sqrt{\frac{3}{2}}$
    (d) $\frac{1}{4}$

31. Proton is
    (a) lepton
    (b) a baryon
    (c) nucleon
    (d) a special kind of meson

32. The hyperfine splitting of the atomic levels is due to
    (a) spin–orbit interaction
    (b) electron spin–nuclear spin interaction
    (c) electron spin–electron spin interaction
    (d) relativistic correction

33. The unit for polarizability of an atom is
    (a) debye/volt
    (b) debye/volt-cm
    (c) volume
    (d) volt/cc

34. Pick out the correct statement from below.
    (a) All the baryons are hadrons.
    (b) Mesons are leptons.
    (c) All the hardons have spin half.
    (d) All the leptons have spin half.

35. A linear array of atoms of mass m are connected with a force constant f. The maximum frequency of the lattice wave that can pass through is given by
    (a) $\gamma_{max} = \left(\dfrac{f}{m}\right)^{\frac{1}{2}}$
    (b) $\gamma_{max} = \pi \left(\dfrac{f}{m}\right)^{\frac{1}{2}}$
    (c) $\gamma_{max} = \dfrac{1}{\pi} \left(\dfrac{m}{f}\right)^{\frac{1}{2}}$
    (d) $\gamma_{max} = \dfrac{1}{\pi} \left(\dfrac{f}{m}\right)^{\frac{1}{2}}$

36. The second-order phase transition is characterized by
    (a) a large value of latent heat
    (b) a decrease in volume
    (c) discontinuity in specific heat
    (d) changes in the composition

37. Kondo effect relates to
    (a) the interaction of conduction electrons among themselves
    (b) interaction of magnetic ions with conduction electrons in a non-magnetic metal

(c) interaction of electrons with the photons in a metal

(d) Mott transition

38. The energy of a system is written as $E = E_1 + E_2$. The corresponding partition function obeys the reaction

(a) $Z = Z_1 + Z_2$

(b) $Z = Z_1 \cdot Z_2$

(c) $Z = \dfrac{Z_1}{Z_2}$

(d) $Z = \ln Z_1 + \ln Z_2$

40. One of the techniques to study the Fermi surface is

(a) nuclear magnetic resonance

(b) infrared absorption

(c) Mossbauer effect

(d) De Haus Van Alphen effect

40. Under suitable conditions, tunnelling of superconducting electron pair from one superconductor to another through a thin insulator takes place. This effect is

(a) Meissener effect

(b) Cooper effect

(c) Joseph effect

(d) Muller effect

41. In relativistic mechanics, the total energy T of a particle is written as

(a) $T^2 = p^2 + m^2 c^4$

(b) $T^2 = p^2 c^2 + m^2 c^2$

(c) $T^2 = p^2 c^2 + m^2 c^4$

(d) $T = p^2 c^2 + m^2 c^4$

42. Tacyons are hypothetical particles whose rest mass is

(a) zero

(b) imaginary

(c) real

(d) 0.5 MeV

43. First excited state energy of the hydrogen atom corresponds to

(a) –13.6 eV

(b) –3.79 eV

(c) –1.51 eV

(d) –0.75 eV

44. The number of nearest neighbours for a body centred cubic lattice is

(a) 4

(b) 8

(c) 12

(d) 16

45. For transition from a state with rotational quantum number j to a state with quantum number $j - 1$, the frequency of the emitted photon is

   (a) $\dfrac{2B}{h}(j-1)$
   (b) $\dfrac{2B}{h}j$
   (c) $\dfrac{4B}{h}$
   (d) $\dfrac{h}{2B}j$

46. In vibrational-rotational spectra, R branch corresponds to

   (a) $\Delta v = 0, \Delta j = 1$
   (b) $\Delta v = 1, \Delta j = -1$
   (c) $\Delta v = 1, \Delta j = 0$
   (d) $\Delta v = 1, \Delta j = 1$

47. Larmer precession corresponds to

   (a) the precession of the vector L around vector J
   (b) the precession of the spin vector S around the vector J
   (c) the precession of the vector J around the applied magnetic vector B
   (d) the precession of the vector L around the applied magnetic field vector B

48. If there are k equations of constraint for a system of N particles then the number of degrees of freedom is

   (a) $3N - k$
   (b) $3N - 3k$
   (c) $N - 3k$
   (d) $3N + k$

49. Canonical transformation implies that

   (a) the form of Hamilton's equations remain unchanged
   (b) the form of Lagrangian equations remain unchanged
   (c) the total energy of the system is conserved
   (d) the generalized moments are no longer constants of motion

50. The differential scattering cross section

   (a) increases with increase in angle of scattering
   (b) increases with increase in impact parameter
   (c) increases with increase in flux density
   (d) decreases with increase in flux density

51. For a system of particles,

   (a) the number of Lagrangian equations is the same as number of Hamilton's equations
   (b) the number of Lagrangian equations is twice the number of Hamilton's equations

(c) the number of Hamilton's equations is twice the number of Lagrange's equation

(d) the number of Lagrangian equations are independent of the number of Hamilton's equations

52. The Bragg's angle for reflection from (111) planes in aluminium (fcc) is 19.2° for an X-ray wavelength of $\lambda = 1.54$ Å. Find out the cubic edge of the unit cell.

(a) 1.01 Å (b) 2.02 Å
(c) 3.03 Å (d) 4.04 Å

53. In a grand canonical ensemble
(a) N, V and E are constants
(b) N is constant but not E
(c) N is allowed to vary but not E
(d) both N and E are allowed to vary

54. For an ideal Bose gas at $T < T_c$, the specific heat at constant volume C_v

(a) varies as $T^3 I^2$ (b) varies as $T^1 I^2$
(c) varies as $T^{-3} I^2$ (d) varies as $T^{-1} I^2$

55. At the Fermi energy the occupation of probability is
(a) 0.5 at $T = 0°K$ (b) 0.25 at $T = 0°K$
(c) 0 at $T = 0°K$ (d) is independent of temperature

56. The parity symmetry of Hamiltonian implies
(a) that the wave functions can be only antisymmetric
(b) that the wave functions can be only symmetric
(c) that the wave functions are either symmetric or antisymmetric
(d) that the wave functions have no symmetry

57. In the occupation number representation, the number operator N is given as

(a) aa^+ (b) $a^+a + \dfrac{1}{2}$ (c) $aa^+ + \dfrac{1}{2}$ (d) a^+a

58. In helium atom triplet states lie between the singlet states. This is due to
    (a) uncertainty principle
    (b) principle of indistinguishability
    (c) spin–orbit interaction
    (d) J–J coupling

59. In partial wave analysis, the effect of scattering results in
    (a) change in energy
    (b) change in angular momentum
    (c) change in the velocity of partial waves
    (d) change in the phase of partial waves

60. The Dirac's relativistic equation
    (a) is a first order differential equation in both space and time
    (b) is a second order differential equation in both space and time
    (c) is a first order differential equation in time but second order in space
    (d) is a first order differential equation in space but second order in time

61. In variational method of approximation
    (a) only ground state wave function can be determined accurately
    (b) one can determine accurately only the ground state energy
    (c) both ground state wave function and energy can be determined accurately
    (d) there are no restrictions as mentioned in the previous statements

62. The number of lattices in an orthorhombic crystal is
    (a) 1 (b) 2 (c) 3 (d) 4

63. A conductor of radius 1 cm carries a current of 10 A. The magnetic field inside the conductor is
    (a) zero
    (b) increases from surface to centre
    (c) decreases from surface to centre
    (d) constant inside the conductor

64. A magnetic dipole is placed in a non-uniform magnetic field. Its motion is described as
    (a) only rotatory
    (b) only translatory
    (c) rotatory and translatory
    (d) no motion

65. An alternating potential is applied to a parallel plate capacitor. Which of the following is true?
    (a) there is displacement current between the plates
    (b) there is conduction current between the plates
    (c) displacement and conduction current between the plates
    (d) neither conduction nor displacement current between the plates

66. A particle is oscillating with displacement x given by $x = A\sin\left(\omega t - \dfrac{\pi}{3}\right)$. The phase of the particle at the starting time is
    (a) $\dfrac{\pi}{3}$
    (b) $-\dfrac{\pi}{3}$
    (c) $\dfrac{\omega\pi}{3}$
    (d) $\dfrac{\pi}{3\omega}$

67. A hollow sphere is filled with water and suspended as a pendulum bob. If the water is allowed to flow slowly through a hole at the bottom, the period of oscillation
    (a) remain the same
    (b) slowly decreases
    (c) slowly increases
    (d) none of the above

68. If the total K.E of all particles before scattering is not equal to that after scattering then it is
    (a) elastic scattering
    (b) inelastic scattering
    (c) Rayleigh scattering
    (d) Newton scattering

69. In a charge free space the following equation is true
    (a) $\nabla \cdot E = \rho$
    (b) $\nabla \cdot D = \rho$
    (c) $\nabla \times E = 0$
    (d) $\nabla \cdot E = 0$

70. If there are p atoms per unit, all the photon dispersion relation will have the following number of acoustical and optical branches
    (a) $(3p - 3)$ optical and $(3p + 3)$ acoustical
    (b) $(3p - 3)$ optical and 3 acoustical
    (c) $(3p - 3)$ acoustical and 3 optical
    (d) $(3p - 3)$ acoustical and $(3p + 3)$ optical

71. The zero point energy of a 3-dimensional oscillator of frequency υ is
    (a) zero
    (b) $h\upsilon$
    (c) $\dfrac{3}{2}h\upsilon$
    (d) $\dfrac{1}{2}h\upsilon$

72. The KE associated with a plane electron wave of wave vector k is given by
    (a) hk
    (b) $\dfrac{1}{2}mk^2$
    (c) $\dfrac{h^2k^2}{8\pi^2 m}$
    (d) $\dfrac{h^2k^2}{m}$

73. The SCR is more useful than the four-layer diode because
    (a) an extra lead is connected to emitter
    (b) an extra lead is connected to collector
    (c) an extra lead is connected to base
    (d) an extra lead is connected to base and emitter

74. Which of the following is the correct statement regarding gauge transformation?
    (a) Potentials are invariant.
    (b) Field vectors are invariant.
    (c) Scalar potential is invariant.
    (d) Any vector is invariant.

75. A coordinate is said to be cyclic if
    (a) the corresponding Hamiltonian is repetitive
    (b) the corresponding Hamiltonian vanishes
    (c) the corresponding momentum vanishes
    (d) none of the above

76. Electric potential at a given place varies as $V = ax^2 + b$. The electric field at the point is given by
    (a) $2ax + b$
    (b) $2ax$
    (c) $2ax - b$
    (d) $-2ax$

77. The decimal number for the hexadecimal 7AF4B is
    (a) 365027   (b) 503627   (c) 350625   (d) 720536

78. Consider the equation $\dfrac{d^2}{dx^2}(\sin 3x) = -9\sin 3x$. The eigen value of the operator $\dfrac{d^2}{dx^2}$ is

(a) 3 (b) 6 (c) 27 (d) −9

79. When diffusion is activated, the dependence of diffusion coefficient on temperature is given by

(a) $\exp\left(\dfrac{E}{kT}\right)$ (b) $\exp\left(\dfrac{-E}{kT}\right)$

(c) $\dfrac{1}{T}$ (d) T

80. The possible values obtained from a measurement of a discrete variable called x are 1, 2, 3, 4. Given that the respective probabilities are $\dfrac{1}{12}, \dfrac{5}{12}, \dfrac{5}{12}, \dfrac{1}{12}$. Then the expectation values of x is

(a) 2.5 (b) 0.5 (c) 1 (d) 10

81. The trace of a matrix is invariant under
(a) orthogonal transformation
(b) similarity transformation
(c) unitary similarity transformation
(d) none of the above

82. The volume of the primitive unit cell, with lattice vectors a, b, c is
(a) $a \times (b \times c)$
(b) $a \cdot (b \times c)$
(c) $a(b \times c)$
(d) $a \cdot b \cdot c$

83. A conservative force F should satisfy the relation
(a) $\nabla \times F = 0$
(b) $\nabla \cdot F = 0$
(c) $\dfrac{dF}{dt} = 0$
(d) $\dfrac{d^2 F}{dt^2}$ is negative

84. Sound waves are always longitudinal. This is true only for media like
(a) solid (b) liquid
(c) gases (d) any medium

85. Thermal expansion of a solid is due to
    (a) increase in the amplitude of harmonic vibrations of atoms
    (b) increase in the amplitude of anharmonic vibrations of atoms
    (c) increase in the concentration of vacancies
    (d) decrease of interatomic forces

86. A Schottky defect in crystal is an example of
    (a) a missing atom   (b) an extra atom
    (c) colour centre   (d) dislocation

87. Electronic specific heat of metals at low temperature varies with temperature T as
    (a) T^1 (b) T^3 (c) T (d) $T^{\frac{3}{2}}$

88. The conductivity of a pure semiconductor is
    (a) proportional to temperature
    (b) inversely proportional to temperature
    (c) increases exponentially with temperature
    (d) decreases exponentially with increasing temperature

89. The coordination number of the rock salt structure is
    (a) 4   (b) 6   (c) 8   (d) 10

90. The eutectic reaction is (S-Solid, L-liquid)
    (a) $S_1 \leftrightarrow S_2 + S_3$ (b) $L \leftrightarrow S_1 + S_2$
    (c) $L_1 + S_1 \leftrightarrow L_2 + S_2$ (d) $L_1 + S_1 \leftrightarrow S_2 + S_3$

91. A polycrystalline material always contains
    (a) crystals of different chemical composition
    (b) crystallites of the same composition but different structure
    (c) crystallites with different orientations
    (d) crystallites of different sizes with same orientation

92. All even–even nuclei have a spin
    (a) $\frac{h}{2}$ to $\frac{9h}{2}$ (b) $I = 0$
    (c) $I < 0$ (d) any value

93. A very powerful tool for nuclear spin determination is
    (a) Raman spectroscopy
    (b) FT-IR spectroscopy
    (c) Mossbauer spectroscopy
    (d) Microwave spectroscopy

94. What happens to the p–n junction of Ge or Si type, if it is doped with the p and n type carriers of the order of 10^{19} atoms/cm^3?
    (a) Width of the depletion region is of the order 10 m.
    (b) Width of the depletion region is of the order 10^{-3} m.
    (c) Width of the depletion region is of the order 10^{-5} m.
    (d) Width of the depletion region is of the order 10^{-10} m.

95. AC compliance of transistor amplifier is
    (a) DC output voltage
    (b) DC output current
    (c) AC output voltage
    (d) AC output current

96. NOR–NOR circuit is equivalent to
    (a) NAND–NOR
    (b) AND–OR
    (c) OR–AND
    (d) NOR–NAND circuits

97. The packing fraction of a face centred cubic system is
    (a) 0.52   (b) 0.68   (c) 0.74   (d) 0.25

98. The number of symmetry elements in D_{2h} point group is
    (a) 2   (b) 4   (c) 6   (d) 8

99. Why don't molecular bonding type solids form above room temperature?
    (a) Binding energy is very high compared to thermal energy at room temperature.
    (b) Binding energy is very low compared to thermal energy at room temperature.
    (c) Binding energy is in order of the thermal energy at room temperature.
    (d) None of the above.

100. Which is the low level computer language?
    (a) BASIC
    (b) FORTRAN
    (c) PASCAL
    (d) ASSEMBLY

ANSWERS

1. (b)	2. (c)	3. (b)	4. (d)	5. (a)
6. (a)	7. (b)	8. (d)	9. (c)	10. (c)
11. (b)	12. (d)	13. (d)	14. (b)	15. (b)
16. (a)	17. (c)	18. (b)	19. (d)	20. (a)
21. (b)	22. (a)	23. (d)	24. (c)	25. (c)
26. (d)	27. (d)	28. (b)	29. (d)	30. (a)
31. (b)	32. (b)	33. (c)	34. (a)	35. (d)
36. (b)	37. (b)	38. (d)	39. (d)	40. (c)
41. (c)	42. (a)	43. (b)	44. (c)	45. (b)
46. (d)	47. (d)	48. (a)	49. (a)	50. (b)
51. (c)	52. (d)	53. (d)	54. (a)	55. (a)
56. (c)	57. (d)	58. (b)	59. (d)	60. (a)
61. (b)	62. (d)	63. (a)	64. (c)	65. (a)
66. (b)	67. (c)	68. (b)	69. (d)	70. (b)
71. (d)	72. (c)	73. (c)	74. (c)	75. (c)
76. (d)	77. (b)	78. (d)	79. (b)	80. (a)
81. (c)	82. (b)	83. (a)	84. (a)	85. (a)
86. (a)	87. (c)	88. (d)	89. (b)	90. (b)
91. (c)	92. (b)	93. (c)	94. (d)	95. (c)
96. (c)	97. (c)	98. (c)	99. (b)	100. (d)

TEST PAPER 2

1. Ball 1 is dropped vertically down from the top of a tower. Ball 2 is thrown horizontally from the tower at the same time.
   (a) Ball 1 reaches the ground earlier.
   (b) Ball 2 reaches the ground earlier.
   (c) The difference in time for the ball to reach the ground depends on the horizontal distance travelled by ball 2.
   (d) Both the balls reach the ground at the same time.

2. Pick the correct statement.
   (a) If a body is not accelerating, there must be no forces acting on it.
   (b) Action–reaction forces never act on the same body.
   (c) Mass of the body depends on its location.
   (d) Action equals reaction only if the body is not accelerating.

3. The dimension of pressure gradient is
   (a) $ML^{-2}T^2$
   (b) MLT^2
   (c) ML^2T^2
   (d) $ML^{-2}T^2$

4. The motion in two dimensions is given by the equation $x = 5t^2 + 2$; $y = 2t^2 + 5$ (t = time). The trajectory of the body is
   (a) parabola
   (b) circle
   (c) straight line
   (d) ellipse

5. A body of mass 0.2 kg is dropped from a height of 80 m. A graph between potential energy and height will be
   (a) a straight line
   (b) parabola
   (c) neither a straight line nor a parabola
   (d) cycloid

6. A ring, a disc, a hollow sphere and a solid sphere are allowed to roll down from the top of a wide inclined plane at the same time.
   (a) The ring will reach the bottom first.
   (b) The solid sphere will reach the bottom first.
   (c) The hollow sphere will reach the bottom first.
   (d) The disc will reach the bottom first.

7. The dimension of angular momentum is
   (a) MLT^1
   (b) $ML^{-1}T^1$
   (c) ML^2T^1
   (d) $ML^{-2}T^2$

8. The time period of a simple pendulum in a satellite
   (a) is unity
   (b) is zero
   (c) depends on the weight of the satellite
   (d) is infinity

9. If no torque acts on a body which is rotating, a decrease in moment of inertia acts in such a way that
   (a) $\frac{1}{2}I\omega^2 = \text{constant}$
   (b) $\frac{I}{\omega} = \text{constant}$
   (c) $I\omega = \text{constant}$
   (d) $I\omega^2 = \text{constant}$

10. If R is the radius of the earth, at what distance above the surface of earth will the value of g become 1% of its value at the surface of the earth?
    (a) $2R$
    (b) $3R$
    (c) $5R$
    (d) $9R$

11. Which of the following has no units?
    (a) Young's modulus
    (b) compressibility
    (c) rigidity modulus
    (d) volume strain

12. A cube of ice floats in a glass of water. When the ice cube melts the water level
    (a) rises
    (b) falls
    (c) remains unchanged
    (d) first rises and then falls

13. Angle of contact for a liquid that does not wet the solid which is in contact with it, is
    (a) zero
    (b) greater than 180°
    (c) between 0° and 90°
    (d) between 90° and 180°

14. If we dip in water, capillary tubes of different radii r, then height h to which water will rise is such that
    (a) $\dfrac{h}{r}$ = constant
    (b) hr = constant
    (c) $\dfrac{h}{r^2}$ = constant
    (d) hr^2 = constant

15. When a spiral spring is extended, the strain involved is
    (a) bulk strain
    (b) linear strain
    (c) shearing strain
    (d) combination of all above strains

16. Two parallel plates contain a thin liquid film between them. The normal force to separate the plate is
    (a) smaller, thinner the film
    (b) greater, thinner the film
    (c) greater, smaller the surface tension
    (d) smaller, greater the area of contact

17. A cylindrical bubble and a spherical bubble from the same liquid have the same radii. Then the excess pressure inside the bubble is
    (a) greater for the cylinder
    (b) is the same for both bubbles
    (c) greater for the sphere
    (d) does not depend on shape of bubbles

18. A venturimeter
    (a) measures the surface tension of a liquid
    (b) bulk modulus of liquids
    (c) rate of flow of liquids
    (d) viscosity of liquids

19. The coefficient of viscosity of hot air is
    (a) same as the coefficient of viscosity of cold air
    (b) smaller than that of cold air
    (c) zero
    (d) greater than that of cold air

20. As temperature of water increases, its viscosity
    (a) decreases
    (b) increases
    (c) remains unchanged
    (d) varies in a complicated manner

21. A sample gas expands from volume V_1 to V_2. In which of the following processes is work done maximum
    (a) isochoric   (b) isothermal
    (c) isobaric   (d) adiabatic

22. During adiabatic expansion the gas cools because
    (a) internal energy decreases
    (b) internal energy increases
    (c) internal energy remains constant
    (d) internal energy varies in a complicated manner

23. A black body is radiating heat at temperature T. If its temperature is doubled, the total energy radiated
    (a) doubles   (b) becomes 4 times
    (c) becomes 8 times   (d) becomes 16 times

24. The units of Stefan's constant are
    (a) $Wm^{-2}K^{-1}$ (b) Wm^2K^4
    (c) $Wm^{-2}K^{-4}$ (d) $Wm^{-2}K^4$

25. A copper disc with a hole at its centre is uniformly heated. The diameter of the hole will
    (a) decrease
    (b) increase
    (c) remain the same
    (d) depend on the mass of the disc

26. At nearly 4°C a given mass of water has maximum
    (a) energy
    (b) volume
    (c) density
    (d) specific heat

27. Pick the correct statement. Liquid helium-II
    (a) has a very low thermal conductivity
    (b) has a higher density than helium I
    (c) has a higher surface tension than helium I
    (d) has a negligible viscosity

28. Adiabatic demagnetization of a substance
    (a) decreases the temperature
    (b) increases the temperature
    (c) increases the volume
    (d) decreases the volume

29. The magnetic susceptibility of a solid K changes with temperature T as
    (a) T^2
    (b) $\dfrac{1}{T^2}$
    (c) $\dfrac{1}{T}$
    (d) T

30. If λ is the thermal conducitivity and σ is the electrical conductivity of a metal at a particular temperature T ($T >$ Debye temperature) then
    (a) $\sigma\lambda^3 = $ constant
    (b) $\dfrac{\lambda}{\sigma} = $ constant
    (c) $\dfrac{\sigma}{\lambda^3} = $ constant
    (d) $\sigma^3 \lambda = $ constant

31. In kinetic theory of gases the most probable speed of the molecules of a gas at temperature T is given as
    (a) $\sqrt{\dfrac{8kT}{m\pi}}$
    (b) $\sqrt{\dfrac{3kT}{m}}$
    (c) $\sqrt{\dfrac{mkT}{3}}$
    (d) $\sqrt{\dfrac{2kT}{m}}$

32. For a linear triatomic molecule the ratio of the specific heats $\dfrac{C_p}{C_v}$ is
    (a) $\dfrac{7}{5}$
    (b) $\dfrac{4}{3}$
    (c) $\dfrac{5}{3}$
    (d) $\dfrac{5}{7}$

33. Newton's law of cooling can be deduced from
    (a) Wien's displacement law
    (b) Stefan–Boltzmann law
    (c) the law of equipartition of energy
    (d) Joule–Thomson effect

34. An electric dipole placed in a non-uniform electric field will experience
    (a) only a force   (b) only a torque
    (c) both torque and force (d) nothing

35. Two equal negative charges ($-q$ each) are fixed at points $[0, -a]$ and $[0, a]$ on the y-axis. A positive charge Q is released from the point $[2a, 0]$ on the x-axis. The charge Q will
    (a) execute simple harmonic motion about the origin
    (b) move to the origin and remain at rest
    (c) move to infinity
    (d) execute oscillatory but not simple harmonic motion

36. An electric cell does 5 Joule of work in carrying 10 coulomb of charge around a closed electric circuit. The emf of the cell is
    (a) 2 volt   (b) 2.5 volt   (c) 3 volt   (d) 0.5 volt

37. Eight drops of mercury of equal radii and possessing equal charges combine to form a big drop. Then the capacitance of the big drop compared to each individual drop is
    (a) 18 times   (b) 14 times
    (d) 2 times   (d) 32 times

38. A hollow metal ball 8 cm in diameter is given a charge -4×10^{-8} C. The potential on the surface of the ball is
    (a) –9000 volt   (b) –900 volt
    (c) –90 volt   (d) zero

39. Two metallic spheres, one hollow and the other solid have the same radii and are charged to the same potential. The charge on the hollow sphere is
    (a) 2 times the solid sphere   (b) half of the solid sphere
    (c) 4 times the solid sphere   (d) same as the solid sphere

40. Twelve wires of equal resistance R are connected to form a cube. The effective resistance between two diagonal ends will be

(a) $\left(\dfrac{6}{5}\right)R$ (b) $\left(\dfrac{5}{6}\right)R$ (c) $3R$ (d) $12R$

41. The deflection in a moving coil galvanometer falls from 50 divisions to 10 divisions when a shunt of 12 ohms is connected across it. The resistance of the galvanometer coil is

(a) 24 ohms (b) 36 ohms (c) 48 ohms (d) 60 ohms

42. The resistance of a moving coil galvanometer is 15 ohms and it gives a full scale deflection for 0.01A. It can be converted into a voltmeter for 5 V full scale deflection by connecting a resistor of

(a) 485 ohm parallel (b) 485 ohm series
(c) 0.05 ohm parallel (d) 500 ohm series

43. A charge of 96 coulomb is passed through a copper voltameter. The number of copper ions liberated is

(a) 3×10^5 (b) 3×10^{20} (c) 3×10^{23} (d) 6×10^{26}

44. The emf in a thermoelectric circuit with one junction at 0°C and the other t°C is given by $E = at + bt^2$. The neutral temperature is

(a) $2ab$ (b) $\dfrac{a}{b}$ (c) $-\dfrac{2a}{b}$ (d) $-\dfrac{a}{2b}$

45. A potential difference of 600 volts is applied across the plates of a parallel plate capacitor, plates being separated by 3 mm. An electron projected vertically parallel to the plates with a velocity 2×10^8 ms^{-1} moves undeflected between the plates. The magnetic field in the region between the condenser plates is

(a) 2×10^6 T (b) 600 T (c) 0.1 T (d) 0.2 T

46. A conducting circular loop of radius r carries a constant current i. It is placed in a uniform magnetic field B such that B is perpendicular to the plane of the loop. The magnetic force acting on the loop is

(a) irB (b) $2\pi irB$ (c) zero (d) πirB

47. Two parallel wires A and B carry currents of 10 and 2 amp respectively in opposite directions. Wire A is infinitely long and

wire B is 2 m long. A and B are separated by a distance 10 cm. Then the force on B is

(a) 8×10^{-5} N
(b) 4×10^{-7} N
(c) 4×10^{-5} N
(d) $4\pi \times 10^{-7}$ N

48. A steel wire of length l has a magnetic moment M. It is then bent into a semicircular arc. The new magnetic moment is

(a) M
(b) $\dfrac{2M}{\pi}$
(c) $\dfrac{M\pi}{l}$
(d) Ml

49. The ratio of magnetic field due to a small bar magnet in the end on position to that of broadside on position for the same distance from it is

(a) $\dfrac{1}{4}$
(b) $\dfrac{1}{2}$
(c) 1
(d) 2

50. The period of oscillation of a bar magnet in a vibration magnetometer is 2 seconds. The time period of oscillation of a bar magnet whose magnetic moment is four times that of the first magnet is

(a) 1 second
(b) 4 seconds
(c) 2 seconds
(d) $\dfrac{1}{2}$ second

51. Induction furnace is based on the heating effect of

(a) magnetic field
(b) eddy current
(c) electric field
(d) electrostatic field

52. The energy stores in an inductance coil is 1 joule when a current 0.01 A is established in it. The self inductance of the coil is

(a) 2.59 H (b) 50 H (c) 25 H (d) 200 H

53. For detecting light intensity we use

(a) photodiode in reverse bias
(b) photodiode in forward bias
(c) LED in reverse bias
(d) LED in forward bias

54. A PN junction diode cannot be used
    (a) as a rectifier
    (b) for converting light energy to electrical energy
    (c) for getting light radiation
    (d) for increasing the amplitude of an AC signal

55. When the conductivity of a semiconductor is only due to the breaking of covalent bonds, the semiconductor is called
    (a) donor   (b) acceptor
    (c) intrinsic   (d) extrinsic

56. A semiconductor is known to have an electron concentration of $8 \times 10^{13}/cm^3$ and hole concentration of $5 \times 10^{12}/cm^3$. The semiconductor is
    (a) P type   (b) N type   (c) intrinsic   (d) insulator

57. An NPN transistor is biased to work as an amplifier. Which of the following statements is not correct?
    (a) The electrons go from base region to collector region.
    (b) The electrons go from emitter region to base region.
    (c) The electrons go from collector region to base region.
    (d) The holes go from base region to emitter region.

58. A common emitter amplifier is operated with a DC supply $V_{cc} = 20$ V. The gain of the amplifier is 100. An AC voltage of 2 V peak to peak is applied at the input. The maximum voltage output peak to peak is
    (a) 0.02 V   (b) 0.0 V   (c) 200 V   (d) 20 V

59. In an amplifier, power level is changed from 8 watts to 16 watts. Equivalent dB gain is
    (a) $2\ dB$ (b) $3\ dB$ (c) $5\ dB$ (d) $6\ dB$

60. A ray of light strikes a piece of glass at an angle of incidence of 60° and the reflected beam is completely plane polarized, the refractive index of glass is
    (a) $\sqrt{3}$ (b) $2\sqrt{3}$ (c) $\sqrt{\frac{3}{2}}$ (d) 2

61. A ray of light is incident on a glass plate at an angle of 60°. What is the refractive index of glass if the reflected and refracted rays are perpendicular to each other?

    (a) sin 60°   (b) cos 60°   (c) cot 60°   (d) tan 60°

62. White light is used to illuminate the two slits in a Young's double slit experiment. The separation between the slits is b and the screen is at a distance d ($\geq b$) from the slits. At a point on the screen directly in front of one of the slits, a certain wavelength is missing. The missing wavelength is

    (a) $\dfrac{\lambda 2b^2}{d}$ (b) $\dfrac{\lambda 3b^2}{d}$ (c) $\dfrac{\lambda 2b^2}{d^2}$ (d) $\dfrac{\lambda 3b}{d^2}$

63. The luminous intensity of light source is 500 candela. The intensity of illumination on a surface placed at a distance of 10 m when light is falling normally is

    (a) 1 lux   (b) 2 lux   (c) 3 lux   (d) 5 lux

64. A fish looking up through the water sees the outside world contained in a circular horizon. If the refractive index of water is $\dfrac{4}{3}$ and the fish is 12 cm below the water surface, what is the radius of the circle?

    (a) $12 \times 3 \times \sqrt{5}$ cm   (b) $12 \times \dfrac{\sqrt{5}}{3}$ cm

    (c) $12 \times 3 \times \sqrt{7}$ cm   (d) $12 \times \dfrac{3}{\sqrt{7}}$ cm

65. A perfectly reflecting mirror has an area of 1 cm². Light energy is allowed to fall on it for one hour at the rate of $10W$ cm^{-2}. The force that acts on the mirror is

    (a) 6.7×10^{-8} N   (b) 2.4×10^{-4} N
    (c) 3.35×10^{-8} N   (d) 1.34×10^{-7} N

66. A lens of power +2 dioptres is placed in contact with a lens of power −1 dioptre. The combination will behave like

    (a) convergent lens of focal length 50 cm
    (b) convergent lens of focal length 100 cm

(c) divergent lens of focal length 50 cm
(d) divergent lens of focal length 100 cm

67. The graph drawn with object distance along the x-axis and the image distance along the y-axis for a convex lens is a
    (a) circle   (b) parabola
    (c) rectangular hyperbola   (d) straight line

68. A thin prism P_1 with angle 4° and made from glass of refractive index 1.54 is combined with another thin prism P_2 made from glass of refractive index 1.72 to produce dispersion without deviation. The angle of the prism P_2 is
    (a) 2.6 degrees   (b) 3 degrees
    (c) 4 degrees   (d) 5.33 degrees

69. To obtain achromatic combination of concave and convex lens, the two lenses chosen should have
    (a) equal powers
    (b) equal refractive indices
    (c) equal dispersive powers
    (d) equal product of their powers and dispersive powers

70. The f-number of a camera is 4.5, this means that the
    (a) aperture of the lens is 4.5
    (b) reciprocal of the focal length is 4.5
    (c) focal length of the lens is 4.5 cm
    (d) ratio of the focal length to aperture is 4.5

71. The momentum of a photon of frequency 10^9 cycles/sec. in kgms^{-1} is
    (a) 6.6×10^{-26} (b) 7.3×10^{-29}
    (c) 2.2×10^{-33} (d) 3.1

72. A radio transmitter operates at a frequency of 880 kHz and a power of 10 kW. The number of photons emitted per second are
    (a) 0.075×10^{-34} (b) 1.71×10^{31}
    (c) 13.27×10^{34} (d) 1.327×10^{34}

73. An electron and a proton are moving with same speed. The ratio of their de Broglie wavelength is
    (a) 1900   (b) 2   (c) 20   (d) 3000

74. The threshold frequency for photoelectric effect corresponds to a wavelength of 5000 Å. Its work function is
    (a) 1 joule   (b) 3×10^{-19} joule
    (c) 4×10^{-19} joule   (d) 2×10^{-19} joule

75. In Millikan's oil drop experiment an oil drop of radius r and charge Q is held in equilibrium between the plates of a charged parallel plate capacitor when the potential difference is V. To keep a drop of radius $2r$ and charge $2Q$ in equilibrium between the plates the potential difference should be
    (a) $8V$ (b) $2V$ (c) V (d) $4V$

76. In astrophysics the Hubble's constant has the dimension
    (a) T^1 (b) L^{-1} (c) T (d) L

77. Angular momentum associated with spin was demonstrated by
    (a) Frank–Hertz experiment
    (b) Davission–Germer experiment
    (c) Stern–Gerlach experiment
    (d) Michelson–Morley experiment

78. If elements of principal quantum number $n > 4$ were not allowed in nature, the number of possible elements will be
    (a) 4   (b) 32   (c) 60   (d) 64

79. Energy levels A, B, C of a certain atom correspond to increasing value of energy $E_A < E_B < E_C$. If $\lambda_1, \lambda_2, \lambda_3$ are wavelengths of radiations corresponding to the transitions C to B, B to A and C to A respectively, which of the following relations is correct?
    (a) $\lambda_1 + \lambda_2 + \lambda_3 = 0$ (b) $\lambda_3 = \dfrac{\lambda_1 \lambda_2}{(\lambda_1 + \lambda_2)}$
    (c) $\lambda_3^2 = \lambda_1^2 + \lambda_2^2$ (d) $\lambda_3 = \lambda_1 + \lambda_2$

80. The velocity of sound in air is given as 350 m/s. A plane travelling at mach-3 has a velocity.
    (a) 116.67 m/s (b) 700 m/s
    (c) 1050 m/s (d) 2100 m/s

81. Pitch of a note is directly proportional to its
    (a) frequency (b) amplitude
    (c) wavelength (d) intensity

82. As a result of interference of waves, energy
    (a) is lost
    (b) is gained
    (c) is transmitted
    (d) remains unchanged as a whole but redistributed

83. In a simple harmonic motion, the maximum acceleration is a and the maximum velocity is b. Its time period is
    (a) $\dfrac{2\pi b}{a}$ (b) $\dfrac{2\pi a}{b}$ (c) $\dfrac{a}{2\pi b}$ (d) $\dfrac{b}{2\pi a}$

84. During a negative β-decay
    (a) an electron which is already present within the nucleus is ejected
    (b) a neutron in the nucleus decays emitting an electron
    (c) a part of the binding energy of the nucleus is converted into an electron
    (d) an atomic electron is ejected

85. Einstein's theory of special relativity
    (a) indicates that magnetism in material can be related to motion of electrons
    (b) shows that magnetic and electrical properties are independent of each other
    (c) has no connection to electricity or magnetism
    (d) deals with only motion of particles

86. Einstein's theory of special relativity involves
    (a) canonical transformation (b) gauge transformation
    (c) Lorentz transformation (d) Galilean transformation

87. If c is the velocity of light and p is the momentum of a photon the energy of a photon is given as
    (a) $\dfrac{hp}{c}$
    (b) $c^2 p^2$
    (c) $\dfrac{p^2}{c^2}$
    (d) cp

88. Binding energy of a deuteron is
    (a) -1.11 MeV
    (b) 2.22 MeV
    (c) 1.11 MeV
    (d) -2.22 MeV

89. If A is the mass number then nuclear volume is proportional
    (a) $A^{\frac{2}{3}}$
    (b) A^3
    (c) A
    (d) A^2

90. Nuclear density is
    (a) independent of A
    (b) proportional to A
    (c) proportional to $\dfrac{1}{A}$
    (d) proportional to $\dfrac{1}{A^3}$

91. Madelung constant for an ionic crystal depends on
    (a) lattice parameter only
    (b) ionic charge and the lattice parameter
    (c) the lattice structure only
    (d) ionic charge only

92. Normalization of a wavefunction over a region implies that
    (a) the particle cannot be in that region
    (b) the particle cannot have a definite momentum in that region
    (c) the particle must have a definite momentum in that region
    (d) the particle must be found in that region

93. For a particle of mass m in an one dimensional box of length L, the energy of the first excited state is given by
    (a) $\dfrac{h^2}{8m^2}$
    (b) $\dfrac{2h^2}{8m^2}$
    (c) $\dfrac{4h^2}{8m^2}$
    (d) $\dfrac{8h^2}{8m^2}$

94. The wave functions of a harmonic oscillator
    (a) can only be symmetric
    (b) can only be antisymmetric

(c) can be symmetric or antisymmetric

(d) have no symmetry properties

95. The term degeneracy in quantum mechanics means

  (a) there are more than one wave functions for the same energy value
  (b) there is only one wave function for one energy value
  (c) there are more than one, energy values for the same wave function
  (d) position and momentum of a particle cannot be determined accurately simultaneously

96. The value of $\dfrac{d\eta}{dx}\eta(x\eta)$ at $x = 0$ is

  (a) 0   (b) n (c) $2n!$ (d) $n!$

97. The solution of the differential equation $\dfrac{d^2y}{dx^2} = y$ which vanishes at both plus and minus infinity is

  (a) e^{-x^2} (b) e^{-x} (c) e^{x^2} (d) non-existent

98. If θ is real the modulus of $e^{2l\theta}$ is always

  (a) > 1 (b) < 1 (c) 0   (d) 1

99. The position vectors of the end points of a diameter of a sphere are $2\hat{I} + 3\hat{J} + 4k$ and $-2\hat{I} + 11\hat{I} + 3k$. The radius of the sphere is

  (a) 5.5   (b) 5.0   (c) 4.5   (d) 4.0

100. Parity is not conserved in

  (a) nuclear fission   (b) nuclear fusion
  (c) strong interactions   (d) electro-weak interactions

ANSWERS

1. (d)	2. (b)	3. (a)	4. (c)	5. (a)
6. (b)	7. (c)	8. (d)	9. (c)	10. (d)
11. (d)	12. (c)	13. (d)	14. (b)	15. (a)
16. (b)	17. (c)	18. (c)	19. (d)	20. (a)

21. (c)	22. (a)	23. (d)	24. (c)	25. (b)
26. (c)	27. (d)	28. (a)	29. (c)	30. (b)
31. (d)	32. (a)	33. (b)	34. (c)	35. (d)
36. (d)	37. (c)	38. (a)	39. (d)	40. (b)
41. (c)	42. (c)	43. (b)	44. (d)	45. (c)
46. (c)	47. (a)	48. (b)	49. (d)	50. (a)
51. (b)	52. (d)	53. (a)	54. (d)	55. (c)
56. (b)	57. (c)	58. (d)	59. (b)	60. (a)
61. (d)	62. (b)	63. (d)	64. (d)	65. (a)
66. (b)	67. (c)	68. (b)	69. (d)	70. (d)
71. (c)	72. (b)	73. (a)	74. (c)	75. (d)
76. (a)	77. (c)	78. (c)	79. (b)	80. (c)
81. (a)	82. (d)	83. (a)	84. (b)	85. (a)
86. (c)	87. (d)	88. (b)	89. (c)	90. (a)
91. (b)	92. (d)	93. (c)	94. (c)	95. (a)
96. (d)	97. (b)	98. (d)	99. (c)	100. (d)

TEST PAPER 3

1. The relation between the phase velocity and group velocity for a non-relativistic free particle is
   (a) $v_p = v_g$
   (b) $v_p = 2v_g$
   (c) $v_p = \sqrt{2}v_g$
   (d) $v_p = \frac{1}{2}v_g$

2. The energy eigen values of a three-dimensional isotropic harmonic oscillator is $\hbar\omega\left(n + \frac{3}{2}\right)$ where $n = 0, 1, 2, \ldots$ etc. The degree of degeneracy of quantum state n is
   (a) $n + 1$
   (b) $\frac{((n+1)^*(n+2))}{2}$
   (c) $\frac{(n+2)}{2}$
   (d) $n + 2$

3. The expression for the following operator $\left(A\frac{d}{dA}\right)^2$ is
   (a) $A\frac{d^2}{dA^2} + A\frac{d}{dA}$
   (b) $A^2\frac{d^2}{dA^2}$
   (c) $A\frac{d}{dA}$
   (d) $A^2\frac{d^2}{dA^2} - A\frac{d}{dA}$

4. The group velocity of electromagnetic waves moving with phase velocity C in a dispersive medium of refractive index n is
   (a) $v_g = \dfrac{C}{\left(n + \omega\dfrac{dn}{d\omega}\right)}$
   (b) $v_g = \dfrac{C}{\left(n - \omega\dfrac{dn}{d\omega}\right)}$
   (c) $v_g = \dfrac{C}{\left(n - \omega\dfrac{dn}{d\omega}\right)^2}$
   (d) $v_g = \dfrac{C}{\left(n + \omega\dfrac{dn}{d\omega}\right)^2}$

5. If the average distance between the sun and the earth is 1.5×10^{11} m then the average solar energy incident on the earth is
   (a) 1.1 Cal/cm²/min
   (b) 2.1 Cal/cm²/min
   (b) 3.1 Cal/cm²/min
   (d) 4.1 Cal/cm²/min

6. If a constant current charges a capacitor then the displacement current is
   (a) $I_d = \dfrac{dV}{dt}$
   (b) $I_d = \dfrac{\partial}{\partial t}(C)$
   (c) $I_d = \dfrac{Cdv}{dt}$
   (d) $I_d = C$

7. Divergence of the field is not zero but its curl is zero. This represents
   (a) static electric field
   (b) static magnetic field
   (c) time varying field
   (d) solenoid field

8. If the lifetime of electronic excited state is 1×10^{-9} s, the uncertainity in the energy of the state is
   (a) 4.1×10^{-6} eV
   (b) 4.1×10^{6} eV
   (c) 1.4×10^{6} eV
   (d) 1.4×10^{-6} eV

9. Which of the following is eigen function of the operator $\dfrac{d^2}{dx^2}$
   (a) $\sin^2 x$
   (b) e^{2x}
   (c) $\cos^2 x$
   (d) $\tan x$

10. The Hamilton operator of a free particle moving in one direction under the influence of zero particle energy is
    (a) $\dfrac{\hbar^2}{2m}\left(\dfrac{d}{dx}\right)$
    (b) $\dfrac{\hbar^2}{2m}\left(\dfrac{d^2}{dx^2}\right)$
    (c) $-\dfrac{\hbar^2}{2m}\left(\dfrac{d^2}{dx^2}\right)$
    (d) $-\left(\dfrac{d^2}{dx^2}\right)$

11. The values of E_o for an electron moving back and forth between potential barriers 10^{-7} cm apart ($m = 9.1 \times 10^{-28}$ g) is
    (a) 6.04×10^{-10} erg
    (b) 6.04×10^{-2} erg
    (c) 6.04×10^{-13} erg
    (d) 6.04×10^{-14} erg

12. Schimidt line gives information about
    (a) magnetic moment of the nucleus
    (b) charge of the nucleus
    (c) quadrupole moment of the nucleus
    (d) size of the nucleus

13. In Gamow–Teller transition, the interaction involved are
    (a) vector and tensor
    (b) tensor and axial vector
    (c) scalar and vector
    (d) tensor and pseudoscalar

14. Which of the following reactions is forbidden by energy conservation?
    (a) $\Sigma^+ \to \pi^0 \mu^+ \upsilon_\mu$
    (b) $K^- + d \to \pi^+ + \Sigma^-$
    (c) $K^+ \to \pi^0 + e^+ + \upsilon_e$
    (d) $\mu^- + p \to Å + \upsilon_\mu$

15. In Schimidt model the magnetic moment is
    (a) equal to zero
    (b) entirely due to odd nucleus
    (c) entirely due to even nucleus
    (d) due to all nucleons present

16. The selection rule for Fermi transition is
    (a) $\Delta I = 0$ (I — nuclear spin)
    (b) $\Delta I = 1$ (I — nuclear spin)
    (c) $\Delta I = -1$ (I — nuclear spin)
    (d) $\Delta I = \pm 1$ (I — nuclear spin)

17. The total scattering cross section of s-wave neutron when the phase shift (δv) is δ_0 is
    (a) directly proportional to $\sin^2 \delta_0$
    (b) directly proportional to $\sin \delta_0$
    (c) inversely proportional to $\sin^2 \delta_0$
    (d) inversely proportional to $\sin \delta_0$

18. When a dielectric sphere is placed in a uniform electric field, the field inside the sphere
    (a) becomes zero
    (b) becomes infinity
    (c) is parallel to the initial field
    (d) is perpendicular to the initial field

19. The length of a rocket ship is 100 m on the ground, when it is in flight its length observed on the ground is 99 m. Its speed is
    (a) 0.0423×10^7 m/s
    (b) 4.23×10^7 m/s
    (c) 0.423×10^7 m/s
    (d) 42.3×10^7 m/s

20. The Hamilton for a charged particle moving in an electromagnetic field is
    (a) $\left(\frac{1}{2}\right) m (p - eA)^2 - e\phi$
    (b) $\left(\frac{1}{2}\right) m (p + eA)^2 - e\phi$
    (c) $\left(\frac{1}{2}\right) m (p + eA)^2 + e\phi$
    (d) $\left(\frac{1}{2}\right) m (p - eA)^2 + e\phi$

21. The particle with a mean proper lifetime of $2\mu s$ moves through the laboratory with a speed of 0.9c. Its lifetime as measured by an observer is
    (a) $0.0458 \mu s$
    (b) $0.458 \mu s$
    (c) $4.58 \mu s$
    (d) $45.8 \mu s$

22. The radius of gyration of a homogeneous rod of constant linear density ρ_n and of length b about an axis perpendicular to the rod and through one end is
    (a) $k = IM$
    (b) $k = (IM)^{\frac{1}{2}}$
    (c) $k = \left(\frac{M}{I}\right)^{\frac{1}{2}}$
    (d) $k = \left(\frac{I}{M}\right)^{\frac{1}{2}}$

23. An experimental formula for the number of hours of sleep a child needs is $S = 135 - \left(\frac{Y}{3}\right)$, where S is the number of hours of sleep

needed and Y is the age of the child in years, according to this formula with each passing year a child needs

(a) $\frac{1}{3}$ hrs less sleep (b) $\frac{1}{3}$ hrs more sleep
(c) 1 hr less sleep (d) 1 hr more sleep

24. If z is the translational partition function for a gas molecule, then its entropy is given by

(a) $Nk\left[\log Z + \frac{3}{2}\right]$ (b) $-NkT \log Z$

(c) $NkT^2\left[\frac{d}{dt}(\log Z)\right]$ (d) $NkT\left[\frac{d}{dv}(\log Z)\right]_T$

25. In canonical ensemble, fluctuations
  (a) do not take place in energy and number of particles of the individual systems
  (b) take place in energy, but not in the number of particles of the individual systems
  (c) take place in the number of particles of the individual systems but not in its energy
  (d) take place in energy as well as the number of particles of the individual systems

26. The degree of degeneracy will be large, when
  (a) particle density is large (b) mass of each particle is large
  (c) temperature is large (d) particle density is low

27. The two particles are distributed among three cells. The ratio of accessible states if they obey to M-B, B-E and F-D statistics is
  (a) 9 : 6 : 3 (b) 3 : 6 : 9 (c) 9 : 6 : 6 (d) 9 : 3 : 3

28. The residue of $\frac{Z}{(Z-a)(Z-b)}$ at infinity is
  (a) $a + 1$ (b) -1 (c) -2 (d) $+4$

29. The Fourier series of an odd function $F(t)$ having period t is a
  (a) Fourier cosine series (b) Fourier integral
  (c) Fourier transform (d) Fourier sine series

30. The principle of least squares does not help us
    (a) to determine the best values of the constant
    (b) to write normal equations
    (c) to determine the form of appropriate curve to fit the given data
    (d) none of the above

31. The magnetic equivalent of Coulomb's law is
    (a) Amperes circuital law   (b) Biot–Savart's law
    (c) Faraday's law   (d) Brewster's law

32. Assuming the power of radiation by sun is 3.8×10^{26} W and its radius 7×10^8 m, the Poynting vector of the surface of the sun
    (a) 1675×10^7 W/m^2 (b) 61.75×10^7 W/m^2
    (c) 6.175×10^7 W/m^2 (d) 617.5×10^7 W/m^2

33. The law of conservation of energy for electromagnetic fields in non-conducting medium is (U = energy density and S = poynting vector)
    (a) $\frac{\partial u}{\partial t} + \nabla \cdot S = 0$ (b) $\frac{\partial u}{\partial t} - \nabla \cdot S = 0$
    (c) $\frac{\partial u}{\partial t} + \nabla \times S = 0$ (d) $\frac{\partial u}{\partial t} - \nabla \times S = 0$

34. The Hertz's potential π of an electric dipole of moment P is
    (a) $\frac{P}{r}$ (b) $\frac{1}{4\pi\varepsilon_0}\left(\frac{P}{r}\right)$
    (c) $\frac{1}{4\pi\varepsilon_0}\left(\frac{r}{P}\right)$ (d) $\frac{1}{4\pi\varepsilon_0}\left(\frac{P}{r}\right)^2$

35. β-Brass (CuZn) is a good example for
    (a) high conductivity
    (b) high resistivity
    (c) order–disorder phenomenon
    (d) low diffusivity

36. The Debye length in a plasma is
    (a) proportional to e, the charge
    (b) inversely proportional to e^2
    (c) very small compared to the size of the plasma
    (d) large compared to the size of the plasma

37. The lowest energy species in a process plasma are
    (a) ions and neutrals back-scattered from the cathode
    (b) main plasma electrons
    (c) process gas neutral atoms and molecules
    (d) secondary electrons accelerated through the cathode sheath

38. The hydromagnetic wave in a plasma under the influence of strong magnetic field propagates
    (a) a frequency equal to the ion cyclotron
    (b) a frequency greater than the ionic gyro frequency
    (c) a frequency much less than that of the ion cyclotron frequency
    (d) a frequency equal to the ion gyro frequency

39. A certain orthorhombic crystal has axial units $a : b : c$ of $0.424 : 1 : 0.367$. The Miller indices of crystal faces whose intercepts are $0.212 : 1 : 0.183$ is
    (a) (212)   (b) (101)   (c) (201)   (d) (112)

40. The diamagnetic susceptibility is independent of
    (a) charge distribution in atoms
    (b) applied field
    (c) temperature
    (d) field applied and temperature

41. The velocity of longitudinal wave in the [111] direction of the cubic crystals is

    (a) $v_1 = \dfrac{1}{3}\left[\dfrac{(C_{11} - 2C_{12} - 4C_{44})}{\rho}\right]^{\frac{1}{2}}$

    (b) $v_1 = \dfrac{1}{3}\left[\dfrac{(C_{11} + 2C_{12} + 4C_{44})}{\rho}\right]^{\frac{1}{2}}$

    (c) $v_1 = \dfrac{1}{3}\left[\dfrac{(C_{11} - 2C_{12} + 4C_{44})}{\rho}\right]^{\frac{1}{2}}$

    (d) $v_1 = \dfrac{1}{3}\left[\dfrac{(C_{11} + 2C_{12} - 4C_{44})}{\rho}\right]^{\frac{1}{2}}$

42. The curve of molar susceptibility vs temperature for MnF_2 shows a peak at 72K. This temperature is called
    (a) Curie temperature
    (b) Weiss temperature
    (c) Langevin temperature
    (d) Neil temperature

43. In the variation method of approximation, the procedure yields
    (a) excited state energy
    (b) different energy values depending on the normalization of the wave function
    (c) the correct wave function of the system
    (d) the upper bound of the lowest state of energy

44. If $J^2 / jm >= \lambda \hbar$ then
    (a) $\lambda = m^2$
    (b) $\lambda < m^2$
    (c) $\lambda = 0$
    (d) $\lambda = m \hbar$

45. Due to a dependent harmonic perturbation, a transition probability proportional to time can arise under certain conditions. One of the statements given below is correct. Pick the correct one.
    (a) The perturbation involves a whole spectrum of frequencies which are very closely spaced.
    (b) The phases of such frequencies are unrelated to each other.
    (c) The magnitude of the perturbations and the spacings of the frequencies are smooth.
    (d) None of the above

46. The operator has the eigen function sin4x and eigen value −16. The operator is
    (a) $4\dfrac{d}{dx}$
    (b) $\dfrac{d^2}{dx^2}$
    (c) $-\dfrac{d^2}{dx^2}$
    (d) $-4\dfrac{d}{dx}$

47. Given the three Pauli spin matrices (anticommuting), the fourth (2×2) matrices which anticommute with these three matrices is
    (a) $\begin{pmatrix} 1 & 0 \\ 0 & 1 \end{pmatrix}$
    (b) $\begin{pmatrix} -1 & i \\ 0 & i \end{pmatrix}$
    (c) $\begin{pmatrix} i & 0 \\ 0 & -i \end{pmatrix}$
    (d) non-existent

48. The Clebsh–Gorden coefficient $\left\langle -\dfrac{1}{2} -\dfrac{1}{2} \middle| 1 -1 \right\rangle$ has the value

(a) −1 (b) 1 (c) 0 (d) $\dfrac{1}{\sqrt{2}}$

49. The operator has an eigen value −25 corresponding to the eigen function sin αx. Then the value of α is

(a) 25 (b) 52 (c) −52 (d) 5

50. The forbidden and allowed energy bands in a crystal is due to the presence of

(a) infinite potential wells in the crystal
(b) finite potential wells in the crystal
(c) finite and periodic potential wells in the crystal
(d) free electron in the crystal

51. Let $\psi(x,y)$ be a normalized wave function corresponding to an infinite one-dimensional system. Here

(a) $\psi^*\psi$ is a conserved quantity

(b) $\int\limits_{-\infty}^{\infty} \psi^*\psi dx$ is conserved quantity

(c) $-\dfrac{\hbar}{2m}\left[\psi^*\dfrac{d\psi}{dx} - \dfrac{\psi d\psi^*}{dx}\right]$ is conserved quantity

(d) ψ and ψ^* are independently conserved

52. A hydrogen atom is placed in an external electric field (ξ). The effect of this in its ground state is

(a) a radius in the energy level by $3e\,\xi a$ where a is the Bohr's radius
(b) a lowering in the energy level by $3e\,\xi a$
(c) a split of spectral lines into three components
(d) no change in the energy levels

53. The angular momentum J satisfies the condition $J \times J$. IHJ. The trace of n of the components J_x, J_y and J_z are

(a) 0, 0, 0 (b) $i, 1, 1$ (c) i, i, i (d) $-i, -i, -i$

54. In the Stark effect of first excited state of hydrogen atom the degeneracy is
   (a) four-fold
   (b) partially lifted
   (c) completely lifted
   (d) not affected

55. Probability density in Dirac theory for a free electron is
   (a) $\psi^*\psi$
   (b) $\psi^+\psi$
   (c) $\alpha\psi\psi^+$
   (d) $\psi\psi^+$

56. In the case of forward scattering, the total static scattering cross section is equal to
   (a) $\left(\dfrac{4\pi}{k}\right) \text{Im} f(0)$
   (b) $\left(\dfrac{4\pi}{k}\right) |f(0)|^2$
   (c) $\left(\dfrac{4\pi}{k}\right) \text{Re} f(0)$
   (d) zero

57. The phase velocity of the matter field is
   (a) twice
   (b) one unit
   (c) one-half
   (d) one quarter

58. The stationary state is that particular state for which the probability distributive function $\psi^*\psi$ is
   (a) zero
   (b) dependent of time
   (c) one
   (d) independent of time

59. For the surface at which there is an infinite potential jump, the wave function is
   (a) zero and the first derivative of ψ is not determined
   (b) zero and the first derivative of ψ is determined
   (c) unity and the first derivative of ψ is determined
   (d) unity and the first derivative of ψ is not determined

60. The eigen values of unitary matrix are
   (a) zero
   (b) real
   (c) unit modulus
   (d) purely imaginary

61. If A be an mn matrix of rank r, then the number of linear independent solutions of systems of equations $ax = 0$ is
   (a) n
   (b) r
   (c) $n + r$
   (d) $n - r$

62. If $g(\omega)$ is the Fourier transform of $f(t)$ the Fourier transform of $f(at)$ is

(a) $g\left(\dfrac{\omega}{a}\right)$ (b) $g\left(\dfrac{a}{\omega}\right)$ (c) $ag\left(\dfrac{\omega}{a}\right)$ (d) $\dfrac{1}{a}g\left(\dfrac{\omega}{a}\right)$

ANSWERS

1. (d)	2. (b)	3. (a)	4. (a)	5. (b)
6. (c)	7. (a)	8. (a)	9. (b)	10. (c)
11. (c)	12. (a)	13. (b)	14. (a)	15. (b)
16. (a)	17. (a)	18. (c)	19. (b)	20. (d)
21. (c)	22. (d)	23. (a)	24. (d)	25. (b)
26. (a)	27. (a)	28. (b)	29. (d)	30. (c)
31. (b)	32. (c)	33. (a)	34. (b)	35. (c)
36. (c)	37. (c)	38. (c)	39. (a)	40. (d)
41. (b)	42. (d)	43. (d)	44. (a)	45. (d)
46. (b)	47. (d)	48. (b)	49. (d)	50. (c)
51. (b)	52. (d)	53. (a)	54. (b)	55. (b)
56. (a)	57. (c)	58. (d)	59. (a)	60. (c)
61. (d)	62. (d)			

CPSIA information can be obtained
at www.ICGtesting.com
Printed in the USA
BVHW051930200122
626720BV00011B/370